Nathan F. Dupuis

Elements of Synthetic Solid Geometry

Nathan F. Dupuis

Elements of Synthetic Solid Geometry

ISBN/EAN: 9783337276331

Printed in Europe, USA, Canada, Australia, Japan

Cover: Foto ©berggeist007 / pixelio.de

More available books at **www.hansebooks.com**

ELEMENTS

OF

SYNTHETIC SOLID GEOMETRY

BY

N. F. DUPUIS, M.A., F.R.S.C.

PROFESSOR OF PURE MATHEMATICS IN THE UNIVERSITY OF QUEEN'S COLLEGE,
KINGSTON, CANADA

New York
MACMILLAN AND CO.
AND LONDON
1893

Norwood Press:
J S. Cushing & Co. — Berwick & Smith.
Boston, Mass., U.S.A.

PREFACE.

————◦◦◦————

THE matter of the present work has, with some variations, been in manuscript for a number of years, and has
formed the subject of an annual course of lectures to
mathematical students by whom the subject has been
well received as one of the most interesting in the earlier
part of a mathematical course.

I have been induced to present the work to the public,
partly, by receiving from a number of Educationists
inquiries as to what work on Solid Geometry I would
recommend as a sequel to my Plane Geometry, and partly,
from the high estimate that I have formed of the value
of the study of synthetic solid geometry as a means of
mental discipline.

To me it seems to exercise not only the purely intel-
lectual powers in the development of its theorems, but
also the imagination in the mental building-up of the
necessary spatial figures, and the eye and the hand in
their representations.

In this work the subject is carried somewhat farther
than is customary in those works in which the subject
of solid geometry is appended to that of plane geometry,

but the extensions thus made are fairly within the scope
of an elementary work, and are highly interesting and
important in themselves as forming valuable aids to the
right understanding of the more transcendental methods.

It appears to me that it is a prevalent custom to lay
too little stress on synthetic methods as soon as plane
geometry is passed, and to hurry the student too rapidly
into the analytic methods. If mathematical knowledge
is all that is required, this may possibly be an advan-
tageous course; but if mental culture is, as it should
be, the chief end in a university education, this customary usage is not the best one.

I have found it convenient to divide the work into
four parts, each of which is further divided into sections.

The first part deals with a consideration of the descriptive properties of lines and planes in space, of the polyhedra, and of the cone, the cylinder, and the sphere.

Here I would feel like apologising for the introduction
of a new term, were it not that I believe that its introduction will be fully justified by a careful perusal of
the work.

Legendre, in his notes to his geometry, proposed to
use the word 'corner' (coin) for the figure formed by
the meeting of two planes, and he considered that the
different polyhedral angles should receive special names
as being geometrical figures of different species. Without

discussing this idea, I have employed the word 'corner' to denote a solid or polyhedral angle of not less than three faces, while I have retained the expression 'dihedral angle' in its usual sense. If a dihedral angle be cut by a plane, this cutting plane necessarily cuts through both faces, and the figure of intersection is a plane angle. Whereas, if any polyhedral angle be cut by a plane which intersects all its faces, the figure of section is not a simple angle, but a polygon. Thus the plane angle and the dihedral have this in common, that they can both be measured by the same kind of angular unit, while the affinities of the polyhedral angle are with the polygon.

Moreover, the trihedral angle is a geometrical function of three plane angles and three dihedral angles, neither of which exists without the other, and every polyhedral angle is a geometrical function or combination of plane and dihedral angles, and these form its elements. Hence I have used the term 'three-faced corner' for 'trihedral angle,' and generally 'n-faced corner' for 'n-hedral angle.' This nomenclature is very convenient; but if any Teacher prefers the older forms, he can readily make the necessary change in language.

The rectangular parallelepiped should certainly be supplied with some convenient name. I have adopted the term 'cuboid,' as proposed by Mr. Hayward, as being both convenient and suggestive.

The second part of the work deals with areal relations, that is, the relations among the areas of squares and rectangles on characteristic line-segments of the prominent spatial figures.

The majority of the results, besides being highly interesting in themselves, form data for subsequent higher work.

The third part is devoted to stereometry and planimetry. In this are developed the principal rules and formulæ for the measurement of volumes and surfaces of the more prominent spatial figures which admit of such measurement, and a special section is given to the consideration of volumes and surfaces generated by moving areas and lines, and to the development of the theorems of Pappus or Guldinus.

The fourth and last part begins with an explanation of the principles of conical or perspective projection. By the application of these principles in projecting a circle into a cone and cutting the cone by a plane, the student is introduced to the conic, and is led to understand its meaning, and the relations of the various conics to one another.

The more common properties of the conics are then easily obtained through a study of the curve as a plane section of a circular cone. The latter half of this part is given to spheric geometry. The spheric figure (triangle and polygon) is considered as the section of a corner by a sphere whose centre is at the apex of the corner. The study of spheric figures is thus brought into line with the study of the corner or solid angle, and the leading properties of the spheric triangle are thus most easily and directly obtained.

The whole work is presented to the younger mathematical reader in the hope that it may prove worthy of his careful attention.

At the close of the work there is a large collection of miscellaneous exercises, many of which, being connected with the subjects of inversion and of polar reciprocation in space, are highly suggestive.

I have to acknowledge my indebtedness to Mr. W. R. Sills for assistance in reading the proof-sheets.

N. F. D.

QUEEN'S COLLEGE,
Oct. 1, 1893.

CONTENTS.

---·◦·---

PART I.

DESCRIPTIVE GEOMETRY.

SECT.   PAGE
1. The Line and the Plane. Intersection. Axial Pencil.
   Generation of Plane. Normal. Cone-circle. Constructions   4
2. Two Planes. Dihedral Angle. Planar and Non-planar
   Figures. Skew Quadrilateral   17
3. Sheaf of Lines or Planes. Solid Angle or Corner.
   Properties of Three-faced Corners or Trihedral Angles. Certain Loci   29
4. Polyhedra. Euler's Theorem $C + F = E + 2$. Tetrahedron. Parallelepiped. Pyramid. Prism. Regular
   Polyhedra. Constructions. Nets   45
5. Cone, Cylinder, and Sphere. Tangent Lines and Planes.
   Conditioned Spheres   61

PART II.

AREAL RELATIONS INVOLVING LINE–SEGMENTS OF SPATIAL FIGURES.

1. Tetrahedron. Parallelepiped. Cuboid. Regular Polyhedra   78
2. Sphere. Radical Plane. Radical Line. Tangent Cone.
   Polar   93

PART III.

STEREOMETRY AND PLANIMETRY.

SECT.   PAGE
1. Polyhedra. Theory of Laminæ. Frustum. Prisma-
    toid. Prismoidal Formula. Applications 100
2. Closed Cone. Frustum. Cylinder. Sphere. Zone and
    Segment 124
3. Special Processes. *A*. Spatial Figures generated by
    Translation of a Plane Figure. Pyramid. Cone.
    Sphere. Groin, etc. *B*. Figures of Revolution.
    C. Mean Centre of Area. Theorems. Guldinus'
    Theorem for Volumes 132
4. Areas of Surfaces. Developable Surfaces. Cone.
    Cylinder. Sphere. Mean Centre of Figure. Guldi-
    nus' Theorem for Surfaces 155

PART IV.

PROJECTIONS AND SECTIONS.

1. Perspective Projection. Various Projections. Vanishing
    Point and Line. Anharmonic Relations. General
    Theorem 165
2. Plane Sections of the Cone. Classification of Conics.
    Degraded Forms. Common Properties. Foci and
    Directrices. Special Study of Ellipse. Theorems of
    Apollonius. The Parabola 175
3. Spheric Sections. General Ideas of Spheric Geometry.
    Spheric Line. Pole and Equator. Lune. Spheric
    Triangle. Polar Triangle. Properties of Spheric
    Triangles. Spherical Excess. Superposability and
    Symmetry. Cases of Ambiguity 197

SOLID OR SPATIAL GEOMETRY.

————∞⦂⊛⦂∞————

PART I.

DESCRIPTIVE GEOMETRY.

1. Solid or Spatial Geometry, or the Geometry of Space, deals with the properties and relations of figures not confined to one plane (P. Art. 19).[1]

The elements of spatial figures are the point, the line, the curve, the plane, and the curved surface. The first four of these are defined in plane geometry (P. Arts. 12, 14, 17); but we repeat here the definition of the plane, as upon that definition several corollaries and other definitions depend.

Def. A *plane* is a surface such that the join of any two arbitrary points in it lies wholly in the surface and coincides with it.

Cor. 1. A line cannot lie partly within a plane and partly without it. For the part within the plane must have at least two points in the plane, and must therefore coincide with the plane throughout its whole extent.

[1] References marked P. are to the Author's 'Geometry of the point, line, and circle in the plane.'

1

Cor. 2. A line not coincident with a given plane meets the plane at only one point.

2. A plane is not necessarily limited in extent; or, in other words, a plane extends to infinity in all its directions. For the plane must be coextensive with every coincident line.

Every plane thus theoretically divides all space into two parts, one lying upon each side of the plane. The use of planes thus considered is common in spherical astronomy.

3. In plane geometry the geometric figure is drawn upon the plane of the paper, which properly represents the plane upon which the figure is supposed to lie. In spatial geometry, however, we have only one plane, that of the paper, to stand for and represent all the planes which may be involved in any spatial figure. This is an unavoidable source of confusion to beginners, as the pictured figures in spatial geometry are not representations of the real figures in the same sense as in plane geometry.

Thus equal line-segments and equal angles in a spatial figure will not, in general, appear as equal segments or equal angles in the pictured representation. So, also, squares and circles in space will not, in general, appear as squares and circles on our single available plane, that of the paper. Properly constructed models simplify matters to a very great extent, and should be employed whenever available. The construction of proper models is, however, always difficult, and often impracticable, and for several reasons they cannot serve all the purposes of a diagram. And hence beginners should accustom themselves to reading and interpreting spatial diagrams. These diagrams can be considered only as an aid to building up the figure in the imagination, and facility in reasoning from such diagrams will depend very largely upon the readiness with which the reasoner can make this imaginary construction. The student is accordingly advised to give some care and patience to the constructing of spatial diagrams.

To represent a plane we usually represent a rectangular segment of the plane, and this generally appears in the diagram as some form of parallelogram.

SECTION 1.

THE LINE AND THE PLANE.

4. Theorem. Two planes which coincide in part coincide altogether.

Proof. The part throughout which the planes coincide must be part of a plane, and must therefore admit of an indefinite number of arbitrary points being taken within it, of which no three are in line. These points taken two and two determine an indefinite number of arbitrary lines which coincide in part with both planes. And the planes thus coinciding (Art. 1. Cor. 1) along an indefinite number of arbitrary lines, coincide altogether, and form virtually but one plane.

Def. An indefinite number of lines can lie in one plane. The totality of these is called a *plane of lines*, although the lines, having only one dimension, do not make up any portion of the plane in which they lie.

5. Theorem. The figure of intersection of two planes is a line.

Proof. Let U and V be two planes, and let A and B be any two points in their figure of intersection. Join A, B by a line. Then, since A and B are two points in U, the join AB lies wholly in U (Art. 1. Def.).

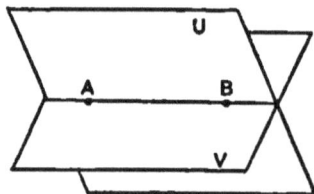

4

For a similar reason the join AB lies wholly in V. Hence it is common to the planes, and is their figure of intersection; and thus the figure of intersection is a line.

Cor. 1. Any number of planes may have one common line. For if they pass through the same two points, A and B, they have the join of A and B as a common line.

Def. A group of planes having one common line is an *axial pencil,* and the line is the axis. In contradistinction to this the pencil of lines in a common plane (P. Art. 203. Def.) is called a *flat pencil.*

Cor. 2. As the line of section of two planes cannot return into itself and form a closed plane figure, so two planes cannot form a closed spatial figure.

6. Theorem. Through any three points not in line,

> 1. One plane can pass.
> 2. Only one plane can pass.

A, B, C are any three points not in line.

1. One plane can pass through A, B, and C.

Proof. Let the plane containing A and B be rotated about the join of A and B.

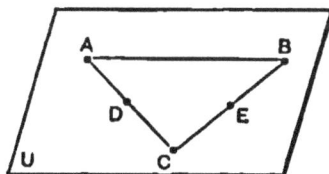

In a complete revolution this plane passes through every point in space, and therefore in some position, U, it passes through C.

2. Only one plane can pass through *A*, *B*, and *C*.

Proof. Take *D*, *E*, any points in the joins *AC* and *BC* respectively. Then *D* and *E* and their join lie in *U*, and in every plane through *A*, *B*, and *C*. Therefore every plane through *A*, *B*, and *C* coincides with *U*, and forms with *U* virtually but one plane.

Cor. 1. Any three points not in line determine a single plane.

Def. Any number of elements so disposed as to lie in one and the same plane are said to be *complanar* or *coplanar*. Thus all the parts of a figure in plane geometry are *complanar*.

Cor. 2. Two intersecting lines are complanar and determine one plane.

For, taking a point in each line, and the point of intersection, we have three points not in line, and the plane through these is the plane of the lines.

Cor. 3. Parallel lines are complanar.

For they have a common point at infinity (P. Art. 220. Def.).

7. Generation of a plane.

L and *M* are any two lines intersecting in *C*, and *N* is a third line intersecting *L* in *B*, and *M* in *A*. Then *L*, *M*, *N* are complanar.

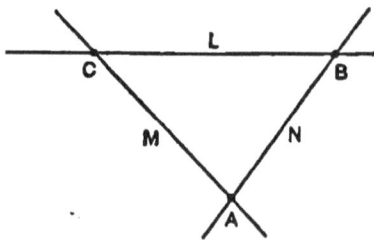

1. When *L* and *M* are fixed and *N* is variable, *N* generates a plane.

Therefore, a plane is generated by a variable line which is guided by two intersecting fixed lines.

Def. The variable line *N* is called the *generator*, and the fixed guiding lines are *directors*.

2. Let *C* go to infinity, and *L* and *M* become parallel.

Therefore, a plane is generated by a variable line guided by two fixed parallel lines.

3. Let the point *A* remain fixed, while *B* moves along *L*.

Then, a plane is generated by a variable line which passes through a fixed point and is guided by a fixed line.

4. Let the point *A* go to infinity; *i.e.* let the generator *N*, fixed in direction only, be guided by the fixed line *L*.

Then, a plane is generated by a variable line having a fixed direction and guided by a fixed line.

8. Theorem. At the point of intersection of any two lines a third line can be perpendicular to both.

AB and *CD* are lines intersecting in *O*. Then some line *OP* is perpendicular to both *AB* and *CD*.

Proof. Let *OP* be ⊥ to *AB*, and let it revolve about *AB* as an axis, *O* being fixed, until it comes into the plane of *AB* and *CD* at *OE* and at *OF*. Then *AB*, *CD*, *EF* are complanar.

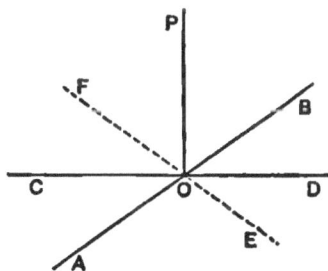

$$\angle AOE = \angle AOP = \mathbf{7}; \qquad \text{(hyp.)}$$

$$\therefore \angle DOE \text{ is } < \text{a} \mathbf{7}.$$

Similarly $\quad \angle BOF = \angle BOP = \mathbf{7};$

and $\quad \therefore \angle DOF \text{ is } > \text{a} \mathbf{7}.$

Therefore, in revolving OP from the position OE to the position OF the $\angle DOP$ changes from less than a right angle at DOE to greater than a right angle at DOF; and hence at some intermediate position OP is $\perp OD$.

Cor. If AB is $\perp$ to CD, and OP is $\perp$ to both, we have three lines mutually perpendicular to each other.

Def. 1. Three concurrent lines mutually perpendicular to one another are called the *three rectangular axes of space*, and their planes are the *rectangular co-ordinate planes of space.* These three lines admit of length measures in three directions, each perpendicular to the other two. Hence, space is said to be of three dimensions, or to contain three dimensions, and it is frequently spoken of as tri-dimensional space, in contradistinction to the two-dimensional space of a single plane, or of plane geometry.

Def. 2. A line lying in a particular plane is a *planar line* of that plane; and when only one plane is under consideration, a planar line will mean a line in that plane.

Def. 3. When OP is perpendicular to both AB and CD, it is perpendicular to the plane which these lines determine (Art. 6. Cor. 2).

OP is then a *normal* to the plane, and O is the *foot* of the normal.

Also, the plane is a *normal plane* to the line OP.

9. Theorem. A normal to a plane is perpendicular to every planar line through the foot of the normal.

OP is ⊥ to *OA* and *OB*, and *OC* is any line through *O* complanar with *OA* and *OB*. Then *OP* is ⊥ to *OC*.

Proof. Take *OA* = *OB* = any convenient length. Join *AB*, cutting *OC* in *C*, and join *PA*, *PB*, *PC*.

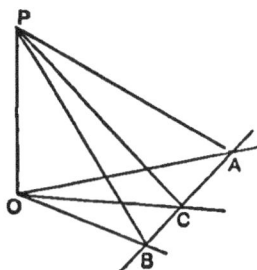

The right-angled triangles *POA* and *POB* are congruent, and therefore *PA* = *PB*. Hence the △ *APB* and *AOB* are each isosceles, and *PC* and *OC* are lines from the vertices to the common base *AB*.

∴ $PB^2 - PC^2 = BC \cdot CA = OB^2 - OC^2$;   (P. Art. 174.)

and   ∴ $PB^2 - OB^2 = PC^2 - OC^2$.

But *POB* being a ⌐,   (hyp.)

$$PB^2 - OB^2 = OP^2 = PC^2 - OC^2.$$

∴ ∠ *POC* is a ⌐.

Cor. 1.   If *O* is fixed while *OA* revolves about *OP* as an axis, *OA* generates a plane to which *OP* is a normal.

Def.   A line is perpendicular to a line which it does not meet when a plane containing one of the lines can have the other as a normal.

Cor. 2.   A normal to a plane is perpendicular to every line in the plane, and all normals to the same plane are parallel to one another.

Cor. 3.   From any point without or within a plane, only ↄ one normal can be drawn to the plane.

10. Theorem.   Of the line-segments from a point without a plane to the plane : —

1. The shortest is along the normal through the point.

2. The feet of equal segments are equally distant from the foot of the normal, and conversely.

3. Of unequal segments, the longer lies further from the normal than the shorter does, and conversely.

P is any point, and PO is normal to the plane U, not passing through P. A, B, C are points in U.

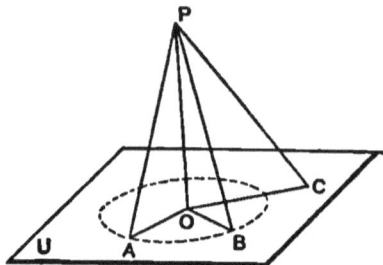

1. PO is $< PA$, A being any point in U other than O.

Proof. $\angle POA$ is a ⌐;   (Art. 9. Cor. 2.)

∴ $\angle PAO$ is acute, and $PO < PA$;   (P. Art. 62.)

and the normal segment PO is the shortest segment from P to the plane U.

2. $PA = PB$; then $OA = OB$.

Proof. The right-angled triangles POA and POB have their hypothenuses equal, and the side PO in common. They are therefore congruent (P. Art. 65), and $OA = OB$.

Conversely, if $OA = OB$, the congruence of the same triangles gives $PA = PB$.

3. PC is $> PA$; then OC is $> OA$.

For the two triangles POA, POC, being each right-angled, give

$$PC^2 = PO^2 + OC^2;$$

and

$$PA^2 = PO^2 + OA^2;$$

$$\therefore PC^2 - PA^2 = OC^2 - OA^2.$$

But $PC > PA$; ∴ $OC > OA$.

And conversely, if $OC > OA$, then $PC > PA$.

Cor. When $PA = PB$, $OA = OB$. Therefore, if PA is of constant length and variable in position, the foot A describes a circle having O as centre and OA as radius. The generation of this circle from a fixed point, P, by a line segment, PA, of constant length, is similar to that of the circle in plane geometry (P. Art. 92), except that in the present case the fixed point is not in the plane of the circle.

Def. 1. The circle described on U with the vector PA, and from the fixed point P, has a relation to the cone, to be considered hereafter, and we shall accordingly call it a *cone circle* to the *vertex P*.

Evidently any circle may be considered as a cone circle, and when so considered, it has an indefinite number of vertices, all lying upon the line which passes through its centre and is normal to its plane.

Def. 2. The distance of a point from a plane is the length of normal intercepted between the point and the plane.

11. *Def.* 1. The *projection* of a point on a plane is the foot of the normal from the point to the plane, and the projection of a line-segment on a plane is the join of the projections of its end-points upon the plane.

It follows, then, that the projection of a line upon a plane which it meets is the planar line which passes through the point where the given line meets the plane, and through the foot of the normal, drawn from any point on the given line to the plane.

Def. 2. The angle between a given line and its projection upon a plane is taken to be the angle between the given line and the plane.

Def. 3.   The angle between two non-complanar lines is the angle between two intersecting lines respectively parallel to the given lines.

12. Theorem.   The angle between a line and its projection on a plane is less than the angle between the given line and any planar line not parallel to the projection.

The line *PO* meets the plane *U* in *O*; *ON* is the projection of *OP* on *U*; *OA* is a line through *O*, parallel to the planar line *L*, which is not parallel to the projection *ON*.

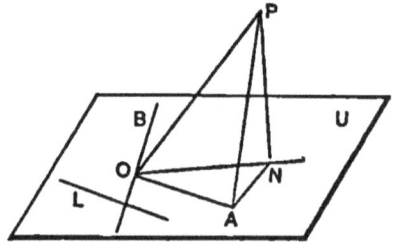

Then   ∠ *PON* is < ∠ *POA*.

Proof.   From *P* draw *PN* perpendicular to *ON*.   *PN* is normal to the plane *U* (Art. 11. Def. 1).

Take *OA* = *ON* and join *PA* and *AN*.

Since   ∠ *PNA* = ⅂, *PA* is > *PN*.

And in the triangles *POA* and *PON*, *PO* is common,

OA = ON, and PA > PN;

∴ ∠ *POA* is > ∠ *PON*.   (P. Art. 67.)

And as *L* is any planar line not parallel to *ON*, the ∠ *PON*, between *PO* and its projection on *U*, is less than that between *PO* and any line in the plane, not parallel to *ON*.

Cor. 1.   Since two intersecting lines make with one another two angles which are supplementary (P. Art.

39), we may say more accurately that the angles, between a line and its projection upon a plane, are the least and the greatest of all the angles made by the given line with lines lying in the plane.

Cor. 2. Since O, P, N are complanar (Art. 6), and $\angle PN\dot{O}$ is a ⌐, the $\angle OPN$ is the complement of the $\angle PON$. Therefore the angle between a line and a plane is the complement of the angle between the line and a normal to the plane.

Cor. 3. Let OB be a planar line $\perp$ to OP.

Since PN is normal to U, OB is $\perp$ to PN (Art. 9. Cor. 2); and hence OB, being $\perp$ to OP and PN, is $\perp$ to ON.

Therefore planar lines which are perpendicular to any line that meets their plane are also perpendicular to the projection of that line upon the plane.

13. *Def.* A line is parallel to a plane when it meets that plane at infinity.

Cor. Any plane through one of two parallel lines is parallel to the other line.

For if L and M be two parallel lines, and the plane U contains L and not M, it can meet M only where L meets M. But L and M meet at infinity (P. Art. 220); therefore M meets U at infinity, or is parallel to U.

SPATIAL CONSTRUCTION.

14. In making constructions in space we assume the ability :

1. To draw through any given point a line parallel to a given line.

2. To pass a plane through any given point or line.

3. To make a plane construction, according to the principles of plane geometry, upon any assumed or determined plane.

Ex. 1. *Problem.* From a given point without a plane to draw a normal to the plane.

Let P be the point, and U be the plane.

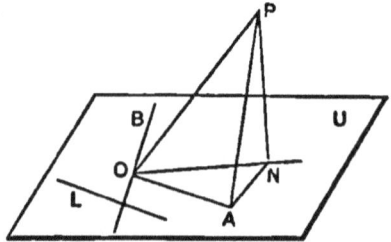

Con. Draw any line OB in U, and from P draw $PO \perp$ to OB (P. Art. 120).

In U draw $ON \perp$ to OB; and from P draw $PN \perp$ to ON. PN is the normal required.

For OB is, by construction, $\perp$ to both OP and ON, and therefore to the plane of these lines, and hence to PN, which lies in this plane (Art. 9. Cor. 2).

Therefore PN is $\perp$ to OB and to ON, and is consequently normal to U.

Ex. 2. *Problem.* To draw a common perpendicular to two non-complanar lines.

Let L, M be the two non-complanar lines.

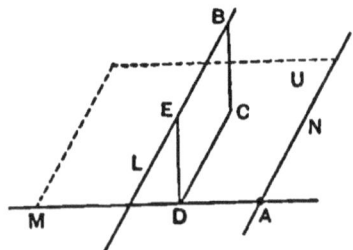

Con. In M take any point, A, and through A draw the line N parallel to L (Art. 14. 1).

M and N determine a plane, U, which is parallel to L. From any point B in L draw BC normal to U (Ex. 1). Then, as L is parallel to U, BC is $\perp$ to L.

Draw CD parallel to L to meet M in D, and from D draw $DE \perp$ to L.

Then DE is $\perp$ to both L and M, or is their common perpendicular.

For DE is $\perp$ to L by construction, and being thus parallel to CB, EC is a rectangle, and ED is normal to U, and therefore $\perp$ to M.

Cor. Since CD can meet M in only one point, only one common perpendicular can be drawn to two non-complanar lines.

EXERCISES A.

1. How many planes at least determine one line ?

2. How many lines at most are determined by 3 planes? by 6 planes ? by n planes ?

3. How many planes at most are determined by 4 points ? by 8 points ? by n points ?

4. Draw a normal to a plane from a point in the plane.

5. Through one of two non-complanar lines, to pass a plane to be parallel with the other line.

6. Show that the common perpendicular to two non-complanar lines is the shortest segment from one line to the other.

7. From a given point in one of two non-complanar lines, to draw a segment of given length to meet the other. The solutions are two, one, or none. Distinguish these cases.

8. Given two non-complanar lines, to draw a segment from one to the other so as to be perpendicular to one of them.

9. Given two non-complanar lines, to draw a segment from one to the other so as to make equal angles with each. Show that this angle may vary from a right angle to the complement of one-half the angle between the given lines.

10. PO meets the plane U (Fig. of Art. 12) at an angle of 30°, and PN is normal to U. OA is a planar line making the angle $POA = 60^\circ$. Show that $\cos AON = \frac{1}{3}\sqrt{3}$.

11. PO meets U at an angle α, and ON is the projection of OP on U. OA is a planar line making the angle $POA = \beta$. Show that $\cos AON = \dfrac{\cos \beta}{\cos \alpha}$.

12. Through the point, where a given line meets a plane, to draw a planar line to make a given angle with the given line.

Examine the limits of possibility.

SECTION 2.

Two Planes — Dihedral Angle — Plane Sections.

15. *Def.* Parallel planes are such as meet only at infinity, *i.e.* which do not meet at any finite point.

Cor. 1. Planes which have a common normal are parallel. For if the planes meet at any finite point, two perpendiculars can be drawn from that point to the same common normal, one in each plane. But this is impossible (P. Art. 61).

Cor. 2. Planes which are not parallel intersect in a line not at infinity. This line is common to the two planes, and is the *common line* of the planes.

When two planes are parallel, their common line is at infinity.

16. *U* and *V* are two planes having *AB* as their common line.

From any point, *P*, in *AB* draw *PC* in *U* and *PD* in *V*, each perpendicular to *AB*.

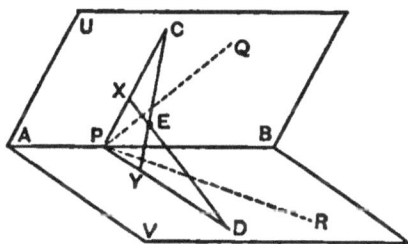

The angle *CPD* is defined as the angle between the planes *U* and *V*. Therefore:

17

Def. 1.   The angle between two planes is the angle between two lines, one in each plane, and both perpendicular to the common line of the planes.

AB is normal to the plane of *PC* and *PD*, and is therefore perpendicular to *CY* and *DX* (Art. 9. Cor. 2).

Hence, if *CY* be ⊥ to *PD*, and *DX* to *PC*, *CY* is normal to *V*, and *DX* is normal to *U*.

And these normals, being complanar, intersect in some point, *E*, and the angle *CED* is the supplement of the angle *CPD*.   Hence, if we consider *CY* and *DX* in the same sense, *i.e.* from distal extremity to foot, or *vice versa*, the angle *CED* is the angle between the normals to the planes, and therefore :

Def. 2.   The angle between two planes is the supplement of the angle between normals to the planes.

When *CP* is perpendicular to *PD*, the planes are perpendicular to one another, and *CP* is normal to *V*, and *DP* to *U*.   Hence :

Def. 3.   Two planes are perpendicular to one another when one of them contains a normal to the other.

17. *Def.* When *PC* is perpendicular to *PD*, and each is perpendicular to *AB*, the three planes *U*, *V*, and the plane of *PCD* are mutually perpendicular to one another.   These planes are then called the *rectangular co-ordinate planes of space*, and the common point, *P*, is the *origin*.

If we assume the positions of these three planes, and therefore the position of the origin, the position of any point in space can be determined by giving its distances from these planes, each distance being affected with a proper algebraic sign. This is the fundamental principle in *analytic* geometry of three dimensions.

18. If PQ be any line in U, and PR be any line in V, meeting the common line AB, in the same point, P, PQ and PR are complanar (Art. 6. Cor. 2); and if W denote their plane, PQ is the common line of U and W, PR is the common line of W and V, and AB is the common line of V and U, and these three lines are concurrent at P.

.Therefore, three planes, no two of which are parallel, and which do not form an axial pencil, determine one point, and this point is the point of concurrence of the three common lines of the planes taken in twos.

This point is at infinity when the three common lines are all parallel.

Cor. Three planes cannot form a closed figure. For the planes determine, at most, three concurrent lines, which, meeting in one common point, can never meet in any other points.

19. *Def.* When a spatial figure, S, is cut by a plane, U, the combination of elements common to S and U form upon U a *plane* figure, which is called the *plane section* of S by U, or simply the section of S by U.

This definition suggests to us a relation existing between plane and spatial geometry.

Plane geometry may be aptly described as a plane section of spatial geometry. The plane upon which the figures of plane geometry lie (P. Art. 11) is the plane of section, and the figures themselves may be considered as sections of spatial figures.

From this connection we may be led to expect that relations existing among plane figures are only particular cases of more general relations existing among spatial figures. And hence we naturally look for many analogies amongst the results of plane and of spatial geometry.

Some of these have appeared already, and others will present themselves in the sequel. And it is worthy of note how often the number two of plane geometry becomes three in spatial geometry. Thus two points determine one line, while three points determine one plane; two lines in the plane determine one point, while it requires three planes to determine one point.

20. The following theorems are self-evident:

    1. The section of a line is a point.

    2. The section of a plane is a line.

Hence spatial figures composed of lines and planes give, in section, plane figures composed of points and lines.

Def. Sections made by parallel planes are parallel sections.

21. Theorem. Parallel sections of a plane are parallel lines.

Proof: If U and U' be parallel planes which cut the plane W, the common lines UW and $U'W$ both lie in W, and as U and U' meet only at infinity (Art. 15), these common lines meet only at infinity and are parallel.

Cor. 1. The section of a system of parallel planes is a system of parallel lines.

Cor. 2. The section of an axial pencil is a set of parallel lines when the section-plane is parallel to the axis; in other cases it is a flat pencil.

22. Theorem. Parallel sections of two intersecting planes contain the same angle.

U and *V* are intersecting planes, and *W* and *X* are two parallel planes of section, the sections being the lines *BA*, *BC, ED,* and *EF.*

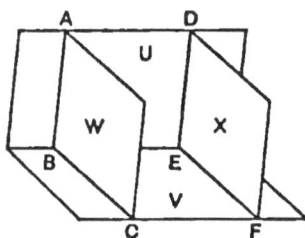

| Then | *AB* is parallel to *DE,* | (Art. 21.) |
| and | *BC* is parallel to *EF.* | |

$$\therefore \angle ABG = \angle DEF.$$

Def. If *W* be normal to the common line of the planes *U* and *V,* the section is called a *right section.* Hence, the angle between two planes is the angle between the two lines which form the right section of the planes.

A system of any number of planes admits of a right section when all the common lines of the planes are parallel. In every case, the term "right section" must have reference to some particular line or set of parallel lines.

DIHEDRAL ANGLE.

23. When we cut two intersecting planes, *U* and *V*, by a third plane, *X*, we get (Fig. of Art. 22),

1. A point *E*, the vertex of the angle *DEF*;

2. The lines *ED* and *EF*, forming the arms of the angle *DEF*.

Now, 1st, to the plane angle *DEF* corresponds the *dihedral* angle between the planes *U* and *V*; and, 2d, to the vertex *E* of the plane angle corresponds the common line, *BE*, of the two planes, this line being called the *edge* of the dihedral angle; and, 3d, to the arms *ED* and *EF* of the plane angle correspond the planes *U* and *V*, called the *faces* of the dihedral angle.

Thus in section a dihedral angle becomes a plane angle, the faces become arms, and the edge becomes the vertex.

If the section be a right section, the plane angle and the dihedral angle have the same measure. And as a plane angle is generated by rotating a line about a point in the line taken as a pole (P. Art. 32), so a dihedral angle is generated by the revolution of a plane about any line in the plane, taken as an axis.

The angular measurements are thus the same for plane and dihedral angles.

24. *Def.* The plane which is normal (Art. 9. Def. 3) to the join of two given points at its middle point, is the right-bisector plane of the join of the points.

Cor. Since a line-segment has only one middle point, and a plane has only one normal at any given point, it follows that a given line-segment has only one right-bisector plane.

A section through the segment gives the segment and its right-bisector, of plane geometry.

25. Theorem. Every point upon the right-bisector plane of a segment is equally distant from the end points of the segment.

Let AB be a given segment, and let U be the right-bisector plane of the segment, passing through its middle point C, and let P be any point on U. Then P is equidistant from A and B.

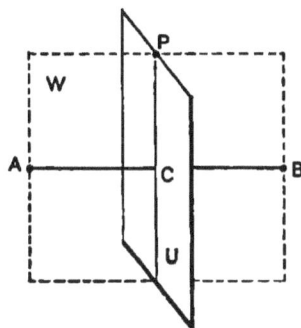

Proof. Since A, B, and P are complanar, let the plane W pass through these points. In the section by W we have the segment AB and its right-bisector CP; and hence $PA = PB$ (P. Art. 53).

It will be here noticed that the proof is obtained immediately by reducing the theorem to depend upon the corresponding one in plane geometry.

In like manner we readily prove the converse:

Every point equidistant from the end points of a given line-segment is upon the right-bisector plane of the segment.

Cor. From this it appears that the locus of a point which is equidistant from two fixed points is the *right-bisector plane* of the join of the points.

26. *Def.* The planes which pass through the edge of a dihedral angle and make equal angles with its faces are the bisectors of the dihedral angle.

The proofs of the following theorems may be obtained at once by making them to depend upon the corresponding theorems in plane geometry.

1. The two bisectors of a dihedral angle are perpendicular to one another.

For proof, make a right section of the dihedral angle and apply (P. Art. 45).

2. Any point upon a bisector of a dihedral angle is equally distant from the faces of the angle.

For proof, make a right section through the point and apply (P. Art. 68).

3. Any point equidistant from the faces of a dihedral angle is on one of the bisectors of the angle.

Proof as in 2.

27. Theorem. Any two lines are divided similarly (P. Art. 201. Def.) by a system of parallel planes.

L and M are two lines cut by the parallel planes U, V, and W. Then L and M are similarly divided.

Proof. A, B, C and A', B', C' are corresponding points of section of the two lines. Through A draw the line N parallel to L, and let it meet the planes at A, P, and Q.

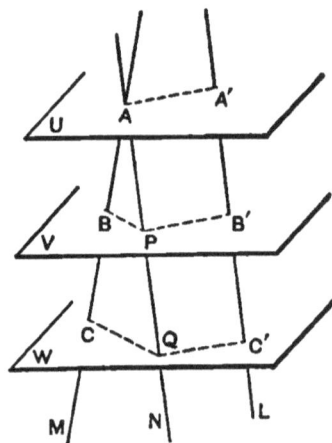

Then L and N being complanar (Art. 6. Cor. 4),

$$AA' \text{ is } \| \text{ to } PB' \text{ is } \| \text{ to } QC';$$

$$\therefore A'B' = AP, \text{ and } B'C' = PQ.$$

But ACQ is a triangle, and BP is parallel to CQ;

$$\therefore\ AB : BC = AP : PQ = A'B' : B'C'.$$

Or the lines L and M are similarly divided.

Cor. 1. The parallel planes of a system divide all lines similarly.

Cor. 2. The segments of parallel lines intercepted between the same two parallel planes are equal.

28. Theorem. If three concurrent non-complanar lines be divided similarly in relation to the point of concurrence, the triplets of corresponding points determine a system of parallel planes.

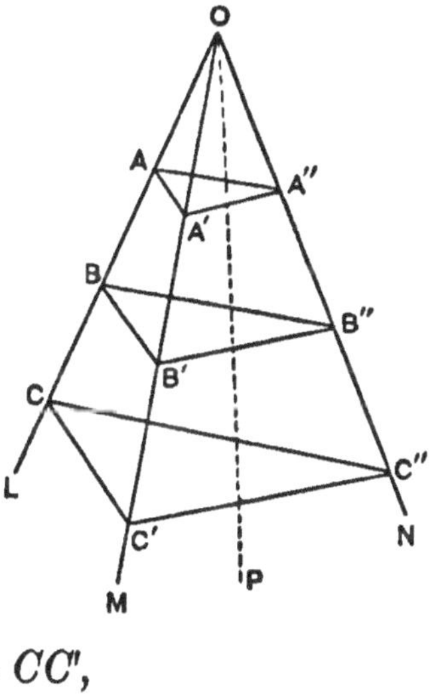

L, M, N are three non-complanar lines concurrent at O, and are divided at A, B, C, $\cdots$, A', B', C', $\cdots$, and A'', B'', C'' $\cdots$, so that

$$OA : AB : BC = OA' : A'B' : B'C'$$
$$= OA'' : A''B'' : B''C''.$$

Then the planes determined by $AA'A''$, $BB'B''$, $CC'C''$, etc., are parallel.

Proof. AA' is $\parallel$ to BB' is $\parallel$ to CC',

and AA'' is $\parallel$ to BB'' is $\parallel$ to CC''. (P. Art. 202. Cor.)

Let OP be normal to the plane $AA'A''$. Then OP is $\perp$ to AA' and AA'' (Art. 9. Cor. 2), and therefore to BB' and BB'', and to CC' and CC''.

Hence OP is normal to the planes $BB'B''$ and $CC'C''$, and the three planes $AA'A''$, $BB'B''$, $CC'C''$ are accordingly parallel (Art. 15. Cor. 1).

Cor. 1. Since AA' is ‖ to BB', and AA'' is ‖ to BB'', etc., the △ $AA'A''$, $BB'B''$, and $CC'C''$ are similar. But the concurrent lines L, M, N determine three planes whose common point is O; therefore parallel sections of three non-parallel planes are similar triangles.

Cor. 2. Since any polygon may be divided into triangles, and similar polygons into similar triangles similarly placed (P. Art. 206), it follows that:
Parallel sections of any number of planes having a common point are similar polygons.

29. *Def.* Four non-complanar lines which intersect two and two in four points, form a *skew-*, or a *gauche-*, or a *spatial* quadrilateral.
The sides of the skew quadrilateral and its two diagonals are six lines connecting four points in space, and form the six edges of a figure, to be described hereafter, called the Tetrahedron.
The skew quadrilateral is a plane quadrilateral with one vertex, and the sides forming it raised out of the plane.

30. Theorem. The joins of the middle points of the opposite sides of a skew quadrilateral bisect one another.
$ABCD$ is a skew quadrilateral, AB and BC lying in a plane different from the plane of CD and DA. AC and BD are the diagonals.

E, F, G, H are middle points of the sides upon which they lie. Then *EG* and *FH* bisect one another.

Proof. EF and *GH* are both parallel to *AC*, and equal to half *AC* (P. Art. 202); they are therefore equal and parallel to one another.

Therefore *EFGH* is a parallelogram, and its diagonals *EG* and *FH* bisect one another (P. Art. 81. 3).

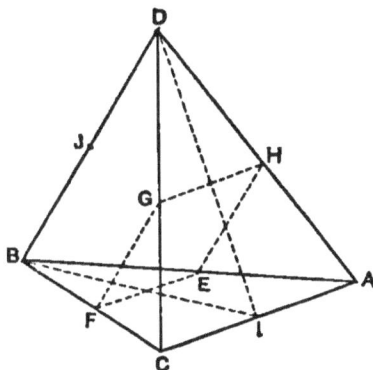

Cor. 1. Let *I* and *J* be the middle points of the diagonals *AC* and *BD*.

Then *ACBD* is a skew quadrilateral, and the joins of middle points of opposite sides are *FH* and *IJ*.

Therefore *FH* and *IJ* bisect one another; and hence *FH, IJ,* and *EG* mutually bisect each other.

Cor. 2. A, B, C, D are four points in space, and *AB, AC, AD, BC, BD,* and *CD* are their six connectors.

Therefore if four points in space be connected two and two by six line-segments, the joins of the middle points of these connectors taken in opposite pairs are concurrent, and mutually bisect one another.

EXERCISES B.

1. Draw a line equally inclined to two intersecting planes. Is the problem definite or indefinite ?

2. If *U* and *V* be two planes, and *U* contains a normal to *V*, show that *V* contains a normal to *U*.

3. Two lines may be drawn, one on each of two intersecting planes, so as to make an angle with one another of any magnitude from zero to a straight angle.

4. If three concurrent non-complanar parallel lines be divided homographically, the planes determined by the triplets of corresponding points, all pass through a common line. When is this line at infinity?

5. If the sides of a skew quadrilateral are equal, the diagonals are perpendicular to one another.

6. What theorem is obtained from 30 by bringing D to the plane of ABC?

7. Draw the shortest path from one point to another so as to touch a given plane in its course, both points being upon the same side of the plane.

8. Show that a skew quadrilateral cannot have four right angles. How many can it have?

9. A, B, C, D are four non-complanar points. Show that the locus of a point which is equidistant from A and B, and also equidistant from C and D, is a line perpendicular to both AB and CD.

10. If A, B, C, D, E, F be any 6 points in space, a point can be found which is equidistant from A and B, equidistant from C and D, and equidistant from E and F.

SECTION 3.

SHEAF OF LINES AND PLANES — SOLID ANGLE OR CORNER.

31. *Def.* Three or more non-complanar lines meeting in a point form a *sheaf of lines*, and three or more planes passing through a common point form a *sheaf of planes*. The common point is in each case called the centre of the sheaf.

The lines and planes which form a sheaf pass through the centre and extend indefinitely outwards from it, but usually we have to consider only those portions which lie upon one side of the centre, and the centre is then commonly called the *vertex* or *apex* of the figure.

In a sheaf of lines the determined planes form a sheaf of planes, and in a sheaf of planes the determined lines f rm a sheaf of lines. So that practically a sheaf of lines and a sheaf of planes are only the same figure differently viewed.

Cor. From Article 27 it follows that the lines of a sheaf are similarly divided by a system of parallel planes.

And from Article 28 it follows that if a sheaf of three lines has its lines similarly divided with reference to the centre, the triplets of corresponding points determine a set of parallel planes.

32. A non-central section of a sheaf of lines and the determined planes is a set of points with their determined lines; and the non-central section of a sheaf of planes and the determined lines is a set of lines with their determined points.

Thus the reciprocity between a sheaf of lines and a sheaf of planes is analogous to that between a set of points and a set of lines in plane geometry.

33. *Def.* If the points in the section of a sheaf of lines be so disposed as to form the vertices of a polygon without re-entrant angles, and only those planes of the sheaf be considered, which, in the section, form the sides of the polygon, the combination of lines and planes in the sheaf forms a *solid angle*, or a *polyhedral angle*, or a *corner*.

L, M, N, K is a sheaf of four lines with centre *O*. Let the sheaf be cut by the plane *U*, giving in section the points *A, B, C, D* corresponding to *L, M, N, K,* respectively. If the polygon *ABCD* is without re-entrant angles, the figure formed by the lines *L, M, N, K,* and the portions of determined planes, *LOM, MON, NOK, KOL,* intercepted between these lines, is a solid angle, or a corner.

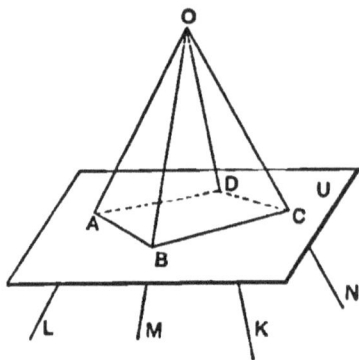

O is the vertex of the corner, *L, M, N, K,* forming the edges or axes of the dihedral angles are its edges; the planes *LOM, MON, NOK,* and *KOL* are its faces; and the angles *LOM, MON, NOK,* and *KOL* are its face-angles.

The term corner or solid angle does not involve any

particular length of line, or extent of plane, or magnitude of angle. It involves the existence of a number of lines forming edges, with the same number of planes limited by these lines and forming faces, and all meeting at a common point to form a vertex.

34. A corner may have any number of faces greater than three, and the same number of edges. The one figured in the preceding article is a four-faced corner, or a tetrahedral angle.

A section of a three-faced corner is a triangle; and as the triangle is the most important of all polygons, so the three-faced corner, or *trihedral angle*, is the most important of all corners.

A corner will be indicated by writing its vertex followed by a point, and then the letters indicating points upon its several edges.

Thus the symbol $O \cdot ABCD$ denotes the four-faced corner as figured in the preceding article.

PROPERTIES OF TRIHEDRAL ANGLES, OR THREE-FACED CORNERS.

35. Theorem. In any three-faced corner the sum of any two face angles is greater than the third.

$O \cdot ABC$ is the three-faced corner.

Proof. If the face angles are all equal to one another, the truth of the theorem is evident. If they are unequal, let the angle LON be $>$ than LOM.

In the plane of L and N draw OK, making the angle $LOK = LOM$, and on M and K take $OB = OD =$ any convenient length, and let A be any point on L, other than O. Let the plane of ABD cut N in C.

Then $\triangle AOB \equiv \triangle AOD.$ (P. Art. 52.)

$\therefore AD = AB$, and $\angle ADB = \angle ABD.$

$\therefore \angle CDB$ is $> \angle CBD$, and CB is $> CD.$ (P. Art. 62. 2.)

But in the $\triangle BOC$ and DOC, $BO = DO$ by construction, OC is common, and $BC > CD.$

$\therefore \angle BOC$ is $> \angle DOC,$ (P. Art. 67.)

and $\because \angle AOB = \angle AOD$ by construction,

$\therefore \angle AOB + \angle BOC$ is $> \angle AOD + \angle DOC.$

Or $\angle AOB + \angle BOC$ is $> \angle AOC.$

36. Problem. To find the locus of a point equidistant from the three edges of a three-faced corner.

O is the vertex, and L, M, N the edges of the three-faced corner.

Let P be a point on the required locus, and PA, PB, PC be perpendiculars upon the edges L, M, and N respectively.

In the right-angled triangles POA, POB, POC, PO is a common hypothenuse, and $PA = PB = PC$ by hypothesis.

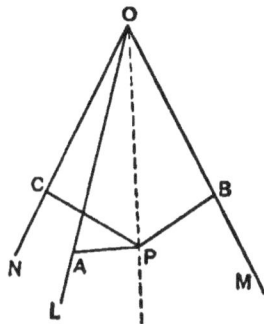

Therefore the triangles are congruent, and $OA = OB = OC$. And the circle through A, B, C is a cone circle with O and P as two vertices.

Therefore OP passes through the centre of this circle and is normal to its plane.

Hence the construction: take $OA = OB = OC$ and join O with the centre of the circle through A, B, and C; this join is the locus required.

Def. The locus just found is a line equally inclined to the three edges, and is an *isoclinal* line to the edges. A plane normal to this line is also equally inclined to the edges and is an isoclinal plane to the edges.

Cor. Since the edges may be considered as indefinite lines extending through the vertex and forming a sheaf of three, the three measures OA, OB, OC may each be taken in two opposite directions, or we can have eight variations of sign in all. But four of these are the other four reversed.

Therefore three lines forming a sheaf have four isoclinal lines and four isoclinal planes through the centre.

37. *Def.* Corners are *equal* when they can be super-imposed so as to form virtually but one corner. In this superposition the vertices coincide, and the edges coincide in pairs, one from each corner.

38. Theorem. Two three-faced corners are equal when the face angles of the one are respectively equal to the face angles of the other, and they are disposed in the same order about the vertices. $O \cdot LMN$ and $O' \cdot L'M'N'$ are two three-faced corners having $\angle LOM = \angle L'O'M'$, $\angle MON = \angle M'O'N'$, $\angle NOL = \angle N'O'L'$, and having these disposed in the same order about the vertices; *i.e.* so that the order of magnitude of the angles is according to the same species of rotation for each. Then the corners are equal.

Proof. Take $OA = OB = OC = O'A' = O'B' = O'C'$, A and A' being on corresponding edges, etc.

The $\triangle AOB \equiv \triangle A'O'B'$, and $AB = A'B'$.

Similarly, $BC = B'C'$, and $CA = C'A'$,

and the $\triangle ABC \equiv \triangle A'B'C'$.

Therefore when $A'B'C'$ is superimposed on ABC, the centres of their circumcircles coincide, and the normals to the planes of these circles at their centres coincide, and hence the vertices of the corners, lying on these normals, coincide (Art. 36), and the two corners, coinciding in all their parts, form virtually but one corner.

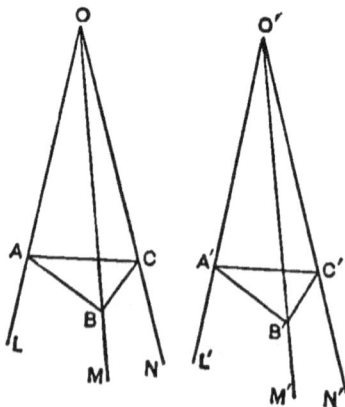

39. Two triangles may be congruent and yet not be superposable until one of them is turned over in the plane. This operation, which is possible and allowable in plane geometry, is not always practicable in spatial geometry.

Suppose the two three-faced corners of the previous article to be so placed that the triangles ABC and $A'B'C'$ lie in one plane, and O and O' are upon the same side of this plane. Then the triangles are directly superposable and the corners are superposable and equal.

But if the triangles ABC and $A'B'C'$ be in the same plane and be directly superposable while O and O' are upon opposite sides of the plane, or if O and O' be upon the same side of the plane while the triangles are not superposable until one of them is turned over in the plane, then the two corners, although having corresponding parts respectively equal, are not superposable, and are not, therefore, equal according to definition.

A little consideration will show that in the non-superposable case, the face angles are disposed in opposite orders about the vertices of the two corners.

Def. Two three-faced corners having corresponding parts respectively equal but not being superposable are said to be *equal by symmetry*, or to be *symmetrical*[1] to each other.

Symmetrical figures are related to each other in the same manner as an object and its image in a plane mirror, or as the right and the left hand; and they

---

[1] The term *conjugate* and *opposable* have both been employed to express the condition here described. But it is obvious that the well-known term *symmetrical* expresses exactly what is meant, and cannot therefore be profitably superseded by any other word.

might be called right-handed and left-handed figures if there were any means of distinguishing between which should be called right-handed, and which left-handed. In certain parts of crystallography the means of distinguishing is apparent, and this terminology is employed.

Two superposable figures can be in perspective with respect to a centre at infinity, while two symmetrical figures can be in perspective with respect to a centre which is the middle point of the joins of corresponding parts.

Cor. It is readily seen that two *n*-faced corners may be superposable and equal, and also that they may be symmetrical and not superposable.

But where there are more than three faces, new possibilities arise, for the face angles may be equal in number and respectively equal in magnitude, and yet the corners may be neither equal nor symmetrical.

40. Theorem. Of two dihedral angles of a three-faced corner and the opposite face angles,

1. The greater face angle is opposite the greater dihedral angle;

2. The greater dihedral angle is opposite the greater face angle.

$O \cdot LMN$ is a three-faced corner having O as vertex, and L, M, N as edges.

From A, any point in L, draw $AB \perp$ to M and $AC \perp$ to N, and from B and C draw, in the plane MN, perpendiculars to M and N respectively, and let these perpendiculars meet in D. Join OD.

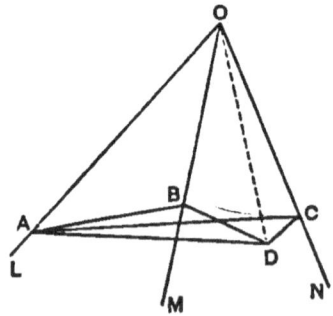

The angles ABD and ACD are respectively the measures of the dihedral angles whose edges are M and N (Art. 16. Def. 1). The △ ADB and ADC are right-angled at D and have AD as a common side, the triangles ABO and ACO are right-angled at B and C, and have AO as a common hypothenuse, and the △ DOB and DOC are right-angled at B and C, and have OD as a common hypothenuse.

1. Let $\angle ABD$ be $> \angle ACD$; then $\angle AOC$ is $> \angle AOB$.

Proof. Since $\angle ABD$ is $> \angle ACD$,

therefore, $\angle BAD$ is $< \angle CAD$, and BD is $< CD$;

and $\therefore BO$ is $> CO$, and AC is $> AB$,

and $\therefore \angle AOC$ is $> \angle AOB$.

2. This, which is the converse of 1, follows from the law of Identity (P. Art. 7).

Cor 1. If a three-faced corner has two dihedral angles equal, it has two face angles equal; and conversely, if it has two face angles equal, it has two dihedral angles equal.

Cor. 2. A three-faced corner with three equal dihedral angles has three equal face angles, and conversely.

Cor. 3. If A, B, C denote the dihedral angles, and a, b, c denote the opposite face angles, the order of magnitude is the same for A, B, C, and a, b, c.

It will be noticed that in this theorem and its corollaries the relations between the dihedral angles and face angles are analogous to those between the angles and sides of a plane triangle.

Def. A three-faced corner with its edges mutually perpendicular to one another is a *rectangular* corner or a *right corner.* It has all its dihedral angles right angles, and all its face angles right angles.

41. Problem. Being given the face angles of a three-faced corner, to construct plane angles which shall have the same measures as the dihedral angles.

$O \cdot LMN$ is the given three-faced corner. To draw a plane angle which shall have the same measure as the dihedral angle whose edge is L.

Constr. Through A, any point in L, draw a plane normal to L, and cutting M and N in B and C.

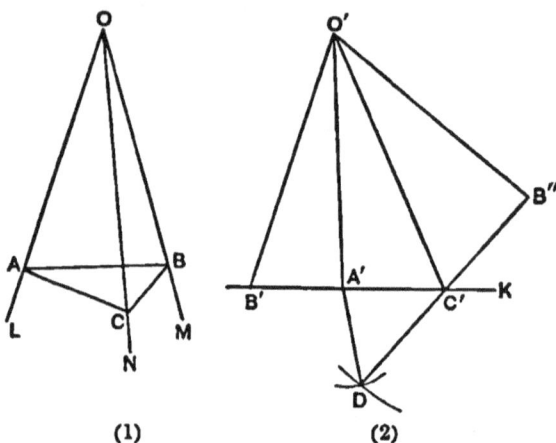

(1)   (2)

In (2) take $O'A' = OA$, and through A' draw a line, K, perpendicular to $O'A'$. Draw $O'B'$, making the $\angle A'O'B' = \angle AOB$, and $O'C'$, making the angle $A'O'C'' = \angle AOC$. Also, draw $O'B'' = O'B'$ and making the $\angle C'O'B'' = COB$. Join $C'B''$.

The $\triangle DA'C'$ constructed with $B'A'$, $A'C'$, and $C'B''$ as sides, has the angle $C'A'D$ equal in measure to the dihedral angle whose edge is L.

Proof.   Since L is normal to the plane of AB and AC, the $\angle BAC$ measures the dihedral angle at L (Art. 16. Def. 1).   And in the construction $O'B' = OB$ and $O'C' = OC$, and hence $A'B' = AB$ and $A'C' = AC$; and also we have made $O'B''C'$ congruent with OBC.

Hence the $\triangle DA'C'$ is congruent with BAC, and the $\angle DA'C'$ measures the dihedral angle at L.

Similarly, the other dihedral angle may be found.

Cor. 1.   Since in the foregoing construction only one triangle is possible with the given elements, the dihedral angles of a three-faced corner are completely given when the face angles are given; and hence the measures of the dihedral angles are expressible in terms of those of the face angles.

Cor. 2.   A three-faced corner is given when its face angles and their order with respect to the vertex are given.

In n-faced corners where n is greater than 3, the giving of the face angles does not determine the dihedral angles, and does not therefore determine the form of the corner. We have the analogue of this in plane geometry, where the giving of the sides of a polygon, with more than three sides, does not determine the form of the polygon.

In general, corners of more than three faces are not of much importance unless they are regular.

Def.   A regular corner has all its face angles equal and all its dihedral angles equal.

42. Theorem.   In any corner the sum of the face angles is less than a circumangle.

Proof. Let the corner have n faces. Cut it by a plane, and we have, as section, a polygon of n sides, the sum of whose internal angles is $2(n-2)$ ⌐s.

Denote, in general, a basal angle of one of the resulting triangular faces by B, and a face angle by F.

At each vertex of the polygonal section, three faces meet to form a three-faced corner, viz. the section itself and two faces of the original corner.

∴ ΣB is $>$ the sum of the internal angles of the section, *i.e.*

$$> 2(n-2) \text{ ⌐s.} \qquad \text{(Art. 35.)}$$

But $$\Sigma B + \Sigma F = 2n \text{ ⌐s.}$$

$$∴ \Sigma F \text{ is } < 4 \text{ ⌐s.}$$

Or the sum of the face angles is less than a circumangle.

43. Let $O \cdot ABC$ be a three-faced corner, and let PS be normal to the plane AOB, PR normal to the plane COA, and PQ normal to the plane BOC.

The angle QPR is the supplement of the dihedral angle at OC, RPS is the supplement of the dihedral angle at OA, and SPQ is the supplement of the dihedral angle at OB (Art. 16. Def. 2).

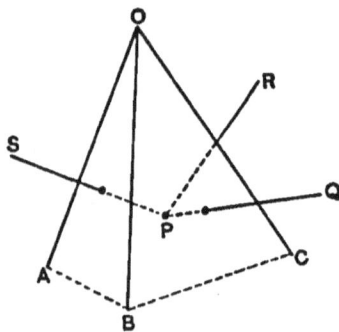

Therefore $P \cdot QRS$ is a three-faced corner in which the face angles are supplementary to the dihedral angles of $O \cdot ABC$.

Also, since OB is normal to the plane SPQ, etc., the face angles of $O \cdot ABC$ are supplementary to the dihedral angles of $P \cdot QRS$.

Similar reasoning will apply to a corner of any number of faces.

Def. Corners so related that the dihedral angles in the one are supplementary to the face angles in the other are called *reciprocal* corners.

Cor. 1. Employing the notation of Art. 40. Cor. 3, for one of the corners, and the letters accented for the other, we have

$$A + a' = B + b' = C + c' = \cdots$$
$$= A' + a = B' + b = C' + c = \cdots = 2\urcorner s.$$

Now, in any corner $a'+b'+c'+\cdots$ is $< 4\urcorner s$; (Art. 42) and $A + B + C + \cdots + a' + b' + c' + \cdots = 2n\urcorner s,$ where n denotes the number of faces.

$\therefore A + B + C + \cdots$ is $> (2n - 4)\urcorner s.$

That is, the sum of the dihedral angles of any corner is greater than the difference between twice as many right angles as the figure has faces, and a circumangle.

Cor. 2. Making $n = 3$, we see that the sum of the dihedral angles of a three-faced corner is greater than two right angles and less than six right angles.

44. Problem. Given the dihedral angles of a three-faced corner, to construct the face angles.

Constr. Take the supplements of the given dihedral angles, and considering these as face angles, construct the corresponding dihedral angles by Art. 41. The supplements of these latter angles are the required face angles.

This construction is evident from the preceding article.

Cor. 1. It is readily seen that only one set of face angles can be obtained when a set of dihedral angles is given; so that when the dihedral angles of a three-faced corner are given, the face angles are given also; and the face angles can be expressed in terms of the dihedral angles.

Cor. 2. A three-faced corner is given when the dihedral angles, and their order, are given.

To construct a three-faced corner when its face angles are given is analogous to constructing a triangle when its sides are given; and to construct the corner when its dihedral angles are given is analogous to constructing the triangle when its angles are given. And this latter is a definite problem with respect to the corner, but an indefinite one with respect to the triangle.

45. Problem. To find the locus of a point equidistant from three given points not in line.

Let *A*, *B*, *C* be the points, and let *U* be the right-bisector plane of *AB*, and *V* be the right-bisector plane of *AC* (Art. 24. Def.).

Every point equidistant from *A* and *B* is on *U* (Art. 25. conv.), and every point equidistant from *A* and *C* is on *V*. And the required locus is the common line of *U* and *V*. But this line evidently passes through the circumcentre of the triangle *ABC* and is normal to its plane.

Hence the locus of a point equidistant from three given points, not in line, is the axis of vertices of the circumcircle of the three points considered as a cone-circle.

Cor. The three right-bisector planes, of the joins of three points, taken two and two, form an axial pencil.

46. Problem. To find a point equidistant from four given points which are not complanar, and no three of which are in line.

Let A, B, C, D be the four points, and let PO be the locus of a point equidistant from A, B, and C. Join D, the fourth point, to any one of the other three, as C, and draw the right-bisector plane, X, of CD.

As D is not complanar with A, B, and C, the plane X is not parallel to PO, and therefore meets PO at some point O. But O is equidistant from A, B, and C, and it is also equidistant from C and D.

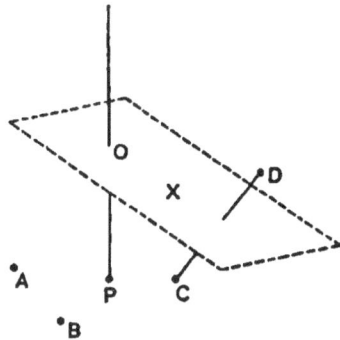

Therefore O is equidistant from A, B, C, and D.

Cor. 1. The line OP is the common line to three bisector planes, namely, those of AB, BC, and CA (Art. 45. Cor.), and X is a fourth plane which goes through the point O. The two remaining bisector planes, those of AD and CD, must pass through the same point O.

Therefore the six right-bisector planes of the joins of four non-complanar points, of which no three are in line, pass through a common point and form a sheaf of planes.

Cor. 2. The four points can be combined to form four different triangles, and the lines, such as PO, which pass through their circumcentres and are normal to their planes, all pass through O and form a sheaf of lines.

Cor. 3. As the line PO can meet the plane X in only one point, there can be only one point equidistant from A, B, C, and D.

EXERCISES C.

1. Any face angle of a three-faced corner is greater than the difference between the other two.

2. Show how to construct a corner symmetrical with a given corner.

3. Show that the three bisectors of the dihedral angles of a three-faced corner have a common line, and that this line is an isoclinal to the three faces.

4. There are four isoclinal lines through the vertex to the three faces of a three-faced corner.

5. In the figure (2) of Art. 41 denote $O'A'$ by p.

Then $A'D = A'B' = p\tan c$; $O'B'' = O'B' = p\sec c$;

$\qquad A'C' \qquad = p\tan b$; $O'C' \qquad = p\sec b$;

and

$C'D^2 = C'B''^2 = DA'^2 + A'C'^2 - 2\,DA' \cdot A'C'\cos A$ (P. Art. 217.)

$\qquad = O'C'^2 + O'B''^2 - 2\,O'C' \cdot O'B''\cos a.$

∴ substituting, and dividing by p^2,

$\qquad \tan^2 c + \tan^2 b - 2\tan c \cdot \tan b \cos A$

$\qquad\qquad = \sec^2 c + \sec^2 b - 2\sec c \cdot \sec b \cos a,$

whence by reduction and dividing by $\cos b \cos c$,

$\qquad \cos a = \cos b \cos c + \sin b \sin c \cos A$;

or, $\qquad \cos A = (\cos a - \cos b \cos c)/\sin b \sin c$;

which expresses a dihedral angle in terms of the face angles.

6. Express a face angle in terms of the dihedral angles. (Employ the property of the reciprocal corner.)

7. If the face angles of a three-faced corner are each 60°, show that the cosine of a dihedral angle is $\tfrac{1}{3}$.

8. In 46 where is the locus if A, B, C are in line ?

9. In 47 where is the point O if the four points be complanar ? where if three points be in line ?

SECTION 4.

POLYHEDRA.

47. *Def.* A spatial figure formed of four or more planes so disposed as to completely enclose a portion of space is a *polyhedron*. It is analogous to the polygon in plane geometry, and its plane section is always some form of polygon.

The *faces* of the polyhedron are those portions of planes which are concerned in forming the closed figure, but for generality the term is sometimes extended to outlying parts of these planes.

The adjacent faces meet by twos to form *edges*, and the edges are concurrent in groups of three or more to form corners.

When a polyhedron is such that no line can meet more than two of its faces, it is *convex*.

48. **Theorem**. In any polyhedron the sum of the number of faces and the number of corners is greater by two than the number of edges.

Proof. Any polyhedron may be supposed to be built up by beginning with one face, and to it adding a second face, and then a third, and so on until the figure is completed.

Denote, in general, the number of corners by C, the number of faces by F, and the number of edges by E.

45

1. Let us start with a single face, U. The number of edges is the same as the number of corners, and we have one face. Therefore the equation $C + F = E + 1$ is satisfied.

2. To U add the face V. In so doing V loses one of its edges, BC, and two of its corners, B and C, by union with similar parts of U. So that in adding V we increase F by 1, and we increase E by one more than the increase of C; and hence the equation $C + F = E + 1$ is still satisfied.

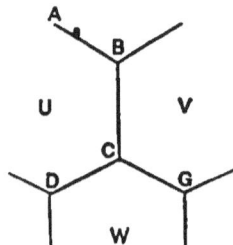

3. To U and V add W. This new face loses two of its edges, DC and CG, and three of its corners, D, C, and G. Here again we add one face and one more edge than corner, so that $C + F = E + 1$ is still satisfied.

4. It is readily seen that in adding any face whatever, that face loses one more corner than edge by union with other faces, until we come to the last face necessary to complete the polyhedron.

This face loses all its edges and all its corners, so that by adding this face we increase the number of faces by 1 without interfering with the numbers of edges or corners. And hence in the completed polyhedron we have

$$C + F = E + 2.$$

This beautiful theorem is usually attributed to Euler, and is known as Euler's theorem on Polyhedra, but it appears to have been known before his time.

CLASSIFICATION OF POLYHEDRA.

49. Polyhedra may be classified as follows:
1. Tetrahedron.
2. Parallelepiped, Cuboid, Cube.
3. Pyramid, Frustum of Pyramid.
4. Prism, Truncated Prism.
5. Prismatoid, Prismoid.
6. The five Regular Polyhedra.
7. A number of Semi-regular Derived Polyhedra.

This classification is not exhaustive, and its divisions are not mutually exclusive. It includes, however, all the polyhedra usually met with.

Polyhedra are not equally important in any sense, and only a few can be said to be important in a descriptive sense.

THE TETRAHEDRON.

50. The three planes which form a three-faced corner, and any fourth plane, not through the vertex, which cuts them all, form the closed figure called a Tetrahedron.

The tetrahedron *ABCD* has four triangular faces, four three-faced corners, and hence four vertices and six edges, *i.e.* the six joins of four non-complanar points no three of which are in line.

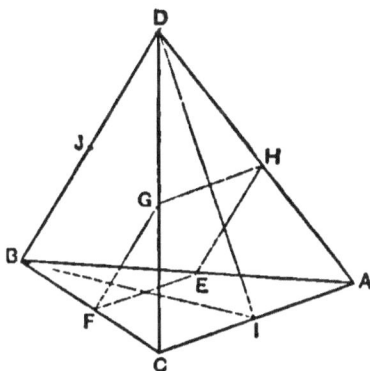

Def. Any face of the tetrahedron may be taken as the *base* of the figure. The three edges which bound the base are then called *basal* edges, and the other three are *lateral* edges.

The joins of the middle points of opposite edges are diameters. There are thus three diameters, *EG*, *FH*, and *IJ*.

51. Theorem. The diameters of a tetrahedron bisect one another.

Proof. ABCD is a skew quadrilateral, and *BD* and *AC* are its diagonals.

But the joins of the middle points of opposite sides of a skew quadrilateral bisect one another (30).

Therefore *EG*, *FH*, and *IJ* bisect one another.

Def. 1. The point of concurrence of the diameters is the *centre* of the tetrahedron. And a section through the centre parallel to a pair of opposite edges is a *middle section*, as *EFGH*.

Cor. There are three middle sections, and these pass through the middle points of the six edges taken in groups of four.

The middle sections are evidently parallelograms, and they intersect by twos along the three diameters.

Def. 2. A *median* of a tetrahedron is the join of a vertex with the centroid (P. Art. 85. Def. 2) of the opposite face.

There are thus four medians, one to each face.

52. Theorem. The medians of a tetrahedron pass through the centre, and are divided at that point so that the part lying between the centre and a face is one-fourth of the whole median to that face.

In the tetrahedron $ABCD$, I is the middle point of BC, and J of AD, and IJ is thus a diameter.

IP is one-third of IA, and P is thus the centroid of the face ABC (P. Art. 85), and DP is the median to the face ABC.

Evidently DP and IJ are complanar, and intersect in some point O. Then O is the centre.

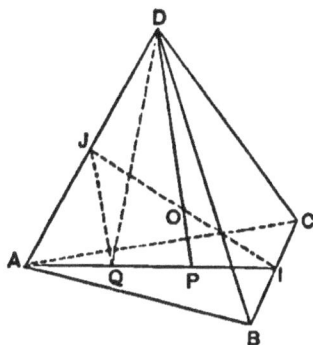

Proof. Draw $JQ \parallel$ to DP to meet IA in Q.

Then, as J is the middle point of AD, so Q is the middle point of AP (P. Art. 84. Cor. 2).

$$\therefore \ AQ = QP = PI = \tfrac{1}{3} AI.$$

And $\because QP = PI$ and OP is $\parallel$ to JQ, O is the middle point of JI, and is therefore the centre (P. Art. 84. Cor. 2). Hence the medians pass through the centre.

Again, $\qquad PO = \tfrac{1}{2} \ QJ$, and $QJ = \tfrac{1}{2} \ PD$.

$$\therefore \ PO = \tfrac{1}{4} PD.$$

Note. — A face of a polyhedron is a segment of a plane, and is in form triangular, rectangular, etc. But in order to avoid such uncouth words as parallelogramic we shall speak of these faces as being triangles, rectangles, parallelograms, etc., although not using these terms strictly as defined in plane geometry. This usage will shorten language and cannot possibly lead to confusion.

The Parallelepiped.

53. *Def.* The *parallelepiped* has six faces, of which each pair of opposite ones are parallel planes.

The contraction ppd. will be frequently used for the word 'parallelepiped.'

Since parallel planes cut any other plane in parallel lines (Art. 21), and since the planes AC and $A'C'$ are parallel and cut the parallel planes AD' and $A'D$, it follows that AB, CD, $A'B'$, and $C'D'$ are all parallel. Similarly, AD, BC, $A'D'$, and $B'C'$ are parallel, and AC', $A'C$, BD', and $B'D$ are parallel.

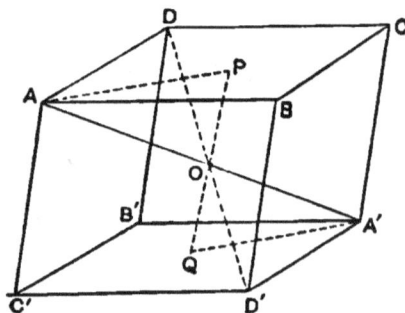

Thus the faces of a ppd. are parallelograms congruent in opposite pairs, and the twelve edges are in parallel sets of four in each set.

The corners, which are eight in number, are each three-faced, and the three edges which meet at any one vertex give the directions of all the edges, and these are therefore called *direction edges*.

Cor. 1. As the $\angle BAD = \angle B'A'D'$, the $\angle DAC' = \angle D'A'C$, and the $\angle BAC' = \angle B'A'C$, the corners having their vertices at A and A' contain face angles which are respectively equal, but these are disposed in opposite orders about the vertices.

The same is true for any other pair of opposite corners.

Therefore, opposite corners of a parallelepiped are symmetrical.

Cor. Considering three-faced corners composed of the same face angles as being of the same variety, there are at most only four varieties of corner in any parallelepiped.

These will be called *representative* corners.

54. By considering the forms of representative corners, all ppds. may be divided into two classes, the *acute* and the *obtuse.*

A denoting any angle, let A' denote its supplement.

Let A, B, C, all acute or all obtuse, be the three face angles at one corner of a ppd.

Then the representative corners are easily seen to be ABC, $AB'C'$, $A'BC'$, and $A'B'C$.

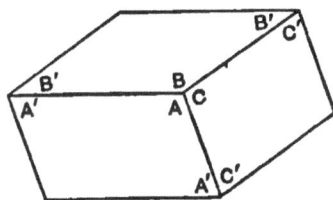

(1) If A, B, C are acute, A', B', C' are obtuse.

Therefore, if a parallelepiped has one corner formed of acute face angles, the other representative corners contain one acute and two obtuse face angles, each.

This is an *acute parallelepiped.*

(2) If A, B, C are obtuse, A', B', C' are acute.

Therefore, if a parallelepiped has one representative corner composed of obtuse face angles, the other representative corners have, each, one obtuse and two acute face angles.

This is an *obtuse parallelepiped.*

It thus appears that no one ppd. can contain all the kinds of corners belonging to ppds.

55. *Def.* The join of opposite vertices in a ppd. is a *diagonal.* These are four in number, viz. AA', BB', CC', and DD' (Fig. of 53).

Since AD is ‖ to BC, is ‖ to $D'A'$ and equal to it, $AD'A'D$ is a parallelogram, and its diagonals bisect one another. Hence AA' and DD' bisect one another; and similarly, AA' and BB' bisect one another, etc.

Therefore, all the diagonals of a ppd. pass through a common point, and are bisected at that point.

The common point of the diagonals is the *centre.*

56. Theorem. Every line-segment passing through the centre of a parallelepiped, and having its end-points upon the figure, is bisected at the centre.

Proof. PQ (Fig. 53) is a line-segment passing through the centre, O, and having its end-points P, Q in the face AC and $A'C'$ respectively.

Join AP and $A'Q$. Then AP and $A'Q$ are complanar, since PQ passes through O; and the plane of AP and $A'Q$ cuts the parallel faces AC and $A'C'$ in parallel lines (Art. 21. Cor. 1).

$$\therefore\ AP \text{ is ‖ to } A'Q.$$

Also, $AO = A'O$, and $\angle AOP = \angle A'OQ$,

and $\angle OAP = \angle OA'Q.$

$$\therefore\ \triangle AOP \equiv \triangle A'OQ,$$

and $OP = OQ.$

Cor. The centre of a ppd. is the centre of every central section.

57. As a parallelepiped has three direction edges, three sections may be made normal to each of these edges respectively. These sections will be forms of the parallelogram.

Def. 1. If none of the sections are rectangles, the ppd. is *triclinic,* and none of its angles, whether face or dihedral, are right angles.

2. If one section is a rectangle, the ppd. is *diclinic,* and four dihedral angles, whose edges are parallel, are right angles.

3. If two sections are rectangles, the ppd. is *monoclinic,* and two sets of four dihedral angles are right angles.

4. If the three sections are rectangles, all the faces are rectangles, and all the dihedral angles are right angles, and all the corners are right corners (Art. 40. Def.). The figure is then a *cuboid.*[1]

Cor. In the cuboid all the diagonals are equal, and the direction lines are mutually perpendicular to one another.

Def. 2. A cuboid with its edges equal is a *cube.* The faces of the cube are squares.

The analogues of the ppd., the cuboid, and the cube, are in plane geometry the parallelogram, the rectangle, and the square.

---

[1] This term was proposed by Mr. Hayward. Before the appearance of Mr. Hayward's work I used the term *orthopiped* for a rectangular parallelepiped. But *cuboid* is evidently a better and a more convenient term.

The Pyramid.

58. *Def.* 1. When a corner of any number of faces is cut by a plane which cuts all the faces, the closed figure so formed is called a *pyramid*.

The cutting plane is the *base*, and the planes which form the corner are faces of the pyramid. The edges which bound the base are *basal* edges, and those which belong to the corner are *lateral* edges. The vertex of the corner is the *vertex* or *apex* of the pyramid.

Def. 2. Pyramids are classified into triangular, square, etc., according to the character of the base. A triangular pyramid is a tetrahedron.

59. *Def.* If a pyramid be cut by a plane parallel to its base, the portion lying between the base and this cutting plane is called a *frustum* of a pyramid.

The frustum has thus two bases, a lower and an upper, or a major base and a minor base.

From Art. 28. Cor. 2, it follows that the two bases of the frustum of a pyramid are similar polygons.

The Prism.

60. When the vertex of a pyramid goes to infinity in a direction normal to the base, the lateral edges become parallel lines, and the resulting figure is not a closed figure. But under like circumstances the frustum becomes a closed figure with two congruent bases, and is called a *prism*.

If one edge of a prism is normal to a base, all the edges are normal, and the lateral faces are rectangles. This is called a *right prism*.

And if one of the lateral edges is inclined to the base, they are all inclined at the same angle. This is an *oblique prism*.

Prisms are usually named from the character of the right section. Thus a right rectangular prism is a cuboid, and a parallelepiped may be a right prism or an oblique prism, depending upon its kind (Art. 57).

THE REGULAR POLYHEDRA.

61. *Def.* A regular polyhedron is one in which all the faces are regular polygons of the same number of sides, and all the corners are formed by the same number of faces.

This implies that all the edges are equal, that all the face-angles are equal, and that all the dihedral angles are equal.

On account of the perfect symmetry of the figure, it must have a definite centre equally distant from each face and equally distant from each vertex. The normal at the centre of each face passes through the centre of the figure, and the line from a vertex to the centre is an isoclinal to the edges of that vertex and to the faces of that vertex.

One of the regular polyhedra is familiarly known as the cube.

62. Theorem. There cannot be more than five regular polyhedra.

Proof. The least number of faces which can form a corner is three, and these must not be complanar. Therefore the three face-angles must together be less than a circumangle, or a face-angle must be less than four-thirds of a right angle (Art. 42).

The only regular polygons having their internal angles less than $\frac{4}{3}$ of a right angle are (P. Art. 133. Cor.) the equilateral triangle, the square, and the regular pentagon; and these alone can form the face of a regular polyhedron.

Equilateral Triangle.

A corner may be formed of 3, 4, or 5 equilateral triangles, and may therefore be three-, four-, or five-faced.

1. The three-faced corner gives the regular tetrahedron, with 4 faces, 4 corners, and 6 edges.

2. The four-faced corner gives the regular octahedron, with 8 faces, 6 corners, and 12 edges.

3. The five-faced corner gives the regular icosahedron, with 20 faces, 12 corners, and 30 edges.

Square.

Only one corner, a three-faced, can be formed by squares.

4. This gives the cube, with 6 faces, 8 corners, and 12 edges.

Regular Pentagon.

Only one corner, a three-faced one, can be formed.

5. This gives the regular dodecahedron, with 12 faces, 20 corners, and 30 edges.

These are the five regular polyhedra.

63. Euler's theorem, Art. 48, gives

$$F + C = E + 2.$$

Now the numbers denoted by F and C are evidently interchangeable, while E remains the same. That is, if we have a given polyhedron, we can form another polyhedron in which the number of corners is the same as the number of faces in the given polyhedron, and the number of faces is the same as that of the corners in the first polyhedron, while the number of edges remains the same in both.

These polyhedra may be called *reciprocals* of each other, as either may be formed from the other by a sort of reciprocation, the changing of points into planes, and planes into points.

If a point be taken in each face of any polyhedron, preferably the centre where there is one, and these points be joined in every way, provided we join only points which lie on adjacent faces, the joins form the edges of a polyhedron which is reciprocal to the original polyhedron.

If the new polyhedron be treated in the same way, we obtain a third polyhedron, which is reciprocal to the second, and is accordingly of the same species as the first.

64. Applying the principles of the preceding article to the regular polyhedra, we readily see that the octahedron is the reciprocal of the cube, and the dodecahedron is the reciprocal of the icosahedron.

The tetrahedron, having the number of its faces and vertices the same, gives another tetrahedron by reciprocation; or the tetrahedron is *self-reciprocal.*

65. Interesting models of all the polyhedra may be made by drawing proper figures on cardboard; then cutting out the entire piece, and cutting half-way through

the remaining lines. The piece of cardboard may now
be folded along these lines to form the intended figure,
and the edges be fastened together with glue.

The figure drawn on the cardboard is called a *net*.

The net for an obtuse parallelepiped is given in the
diagram. The faces are denoted by *U, V,* and *W,* those

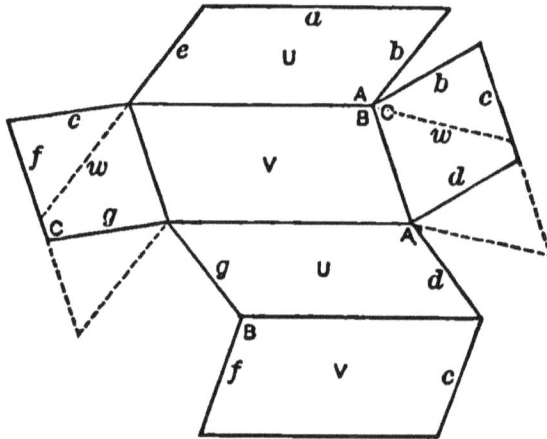

having the same letter being opposite, and therefore congruent parallelograms. The edges which come together
are denoted by the same small letter. Those having
the same letter attached must, of course, be the same
in length. The three obtuse angles concerned are
denoted by *A, B,* and *C.* All the other angles are then
known.

If the angle *C* were acute, as indicated by the dotted
lines, the ppd. would be acute. And the same results
would be obtained by making either *A* or *B* acute.

As the net is drawn, the ppd. will be triclinic. If the
U faces be rectangles, the ppd. will be diclinic; if both
U and *V* are rectangles, it will be monoclinic; and if all

the faces be rectangles, the figure will be the cuboid; and if all squares, the cube.

The accompanying diagrams give nets for the regular polyhedra other than the cube.

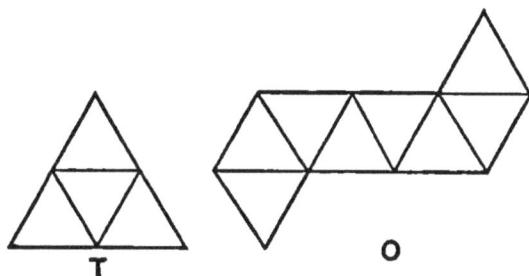

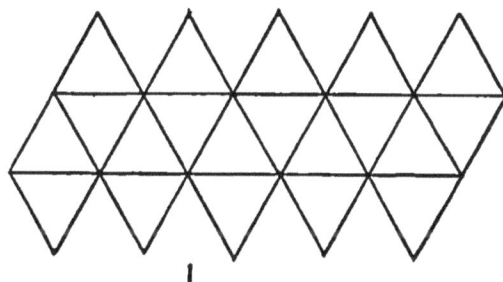

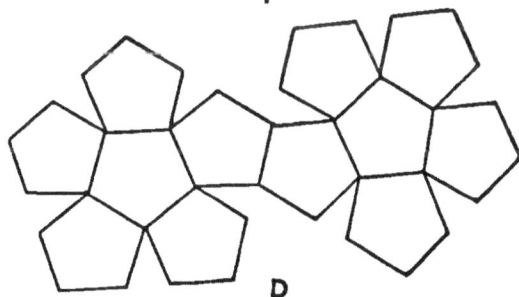

Nets for prisms and pyramids and frusta need no description.

EXERCISES D.

1. The faces of a polyhedron are 3 squares and 2 triangles. Find the number of edges and of corners and classify the figure.

2. If an n-hedron has all its faces triangles, the number of its corners is $\frac{1}{2}(n+4)$.

3. If P, Q, R, S be the centroids of a tetrahedron, the reciprocal having P, Q, R, S as vertices has the same centre as the original. Also the diameters and medians of the two tetrahedra coincide, except in length.

4. In the regular tetrahedron the diameters are perpendicular to one another.

5. If the diameters of a tetrahedron terminate in the centres of the faces of a cube, then the edges are diagonals of the faces. Thence show how the cube may be transformed into a regular tetrahedron.

6. If AA', BB', CC', and DD' are diagonals of a cuboid, show that the middle points of AB, BC, CA', $A'B'$, $B'C'$, and $C'A$ are complanar.
Find the form of the section through these points.

7. The join of A' with the middle point of AB, and the join of C' with the middle point of BC, divide each other into parts which are as 2 to 1 (Ex. 6).

8. The centres of the adjacent faces of a ppd. are joined. What closed figure is formed? Describe its characteristics.

SECTION 5.

THE CONE, THE CYLINDER, AND THE SPHERE.

66. The three figures here mentioned are the simplest spatial figures having curved surfaces, and they are frequently spoken of as the *three round bodies.*

The cone and the cylinder can be generated by the motion of a straight line, and they are consequently called *ruled surfaces.*

The sphere is not a ruled surface, but a surface of *double curvature.*

Def. A surface which can be generated by the revolution of a plane figure about an axial line in its plane, is a *surface of revolution.*

The sphere is a surface of revolution.

The cone and the cylinder may or may not be surfaces of revolution.

Solid geometry furnishes other interesting examples of ruled surfaces besides the cone and cylinder, and of surfaces of revolution besides the sphere. As examples of the first we have the common conoid, the hyperboloid of one sheet, and the elliptic paraboloid; and of the second, the oblate spheroid, the prolate spheroid, and the anchor ring.

THE CONE.

67. *Def.* 1. In general, a variable line which passes through a fixed point and is guided by a fixed plane curve, not complanar with the point, generates a *cone*, or has a cone as its locus.

O is a fixed point, and *APB* is a fixed curve not complanar with the point.   The variable line *L* passes through *O*, and meets the curve *APB*. Then *L* generates a cone.

Cor. Since *L* is unlimited in length, the cone extends indefinitely outwards upon both sides of *O*, and is not a closed figure.

Def. 2. *O* is the centre of the cone, and the two parts into which it divides the cone are called the two *nappes* or *sheets* of the cone.

The fixed curve *APB* is the *director*, and the line *L* is the *generator* of the cone.

Any line which coincides with the generator in any of its positions is called a generating line.

Thus every line passing through *O* and lying on the conical surface is a generating line.

68. The director may be any form of curve. If it becomes a line, the cone degrades into a plane (Art. 7. 3); and if the director becomes a point, the cone becomes the line through that point and the centre.

Thus the line and the plane may be looked upon as limiting forms of the cone.

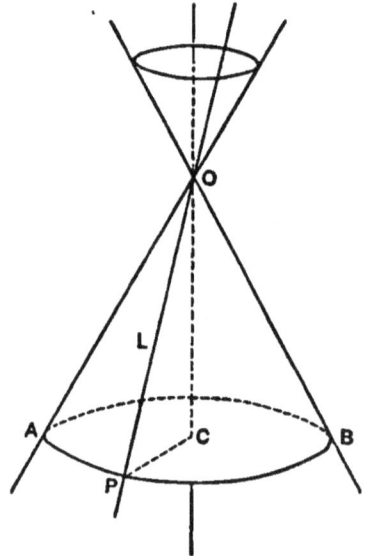

69. When the director is a circle, and the centre O is a vertex to that circle as a cone-circle (Art. 10. Def. 1), the cone is a *right circular cone*, and the line through the centre of the circle and the centre of the cone is the *axis* of the cone.

The circular cone is a figure of revolution, and is the most important of all cones.

The word 'cone' as hereafter employed will mean a right circular cone, unless otherwise qualified.

70. Let C (Fig. of 67) be the centre of the circular director APB. Then CP is constant, and CO is constant, and OCP is a ⌐. Therefore the $\angle POC$ is constant. This angle is the *semi-vertical angle* of the cone.

Hence a circular cone is generated by a line which revolves about a fixed axial line while meeting the latter in a fixed point and at a fixed angle.

Cor. 1. Every section of a circular cone, normal to the axis, is a circle.

Cor. 2. Every section of a circular cone, through the axis, is two lines intersecting at a fixed angle the vertical angle of the cone.

Cor. 3. Every section of a circular cone through the centre is two lines; for the·plane meets the cone along two generating lines.

Cor. 4. Any point on the axis of a circular cone is equidistant from the surface on all sides, and the axis is thus an isoclinal line to the surface.

71. Theorem. Only two generating lines of a cone are complanar.

Proof. Since the generating lines all pass through O, any two of them are complanar.

Let any two particular generating lines meet the director circle in A and P. The plane of these lines meets the plane of the circle in a line (5), and as a line can meet a circle in only two points (P. Art. 94), the plane of OA and OP has only two points coincident with the circle, and therefore only two generating lines lie in this plane.

72. Theorem. A line which is not a generating line can meet a cone in only two points.

Proof. Let M be the line, not passing through O; and let the plane U pass through O, and contain M. If U cuts the cone, it contains two generating lines; and since it contains M, the two generating lines are complanar with M, and meet it in two points, and in only two points; and these points are common to M and to the cone.

Therefore, the line M can meet the cone in two, and in only two, points.

73. If the two points in which a line M, which is not a generating line, meets a cone become coincident, the line becomes a *tangent* line to the cone, and has one point only, a double point (P. Art. 109. Def. 2) in common with the cone.

The plane determined by a tangent line and the generating line through its point of contact is a tangent plane to the cone, and touches the cone along this generating line, which, as it represents the union of two lines, is a double line.

Cor. 1. Evidently all tangent planes to a cone pass through the centre and form a sheaf of planes.

Cor. 2. All tangent planes to a cone intersect one another in lines which pass through the centre and form a sheaf of lines.

74. *Def.* A line through the centre perpendicular to a generating line of a cone generates a second cone, which is the *reciprocal* of the first.

When the vertical angle of the cone is a right angle, these two cones become coincident, and form but one cone.

THE CYLINDER.

75. When the centre of a cone goes to infinity in the direction of the axis, and the director curve remains finite, the cone becomes a *cylinder*, and the axis of the cone becomes the axis of the cylinder. Hence:

Def. 1. A cylinder is the locus of a line which keeps a fixed direction and meets a fixed plane curve which is not complanar with the line.

Def. 2. A circular cylinder is generated by one of a pair of parallel lines while revolving at a fixed distance about the other parallel as a fixed axial line. The fixed line is the axis of the cylinder.

Cor. 1. The cylinder, as defined, is not a closed figure.

Cor. 2. A line can meet a circular cylinder twice, and only twice.

Cor. 3. Sections of a circular cylinder normal to the axis are equal circles.

THE SPHERE.

76. *Def.* A *sphere* is the locus of a semicircle which revolves about its limiting centre line as an axial line.

BAD is a semicircle, and *AB* is its limiting diameter. When *ADB* revolves about *AB* as an axis, the semicircle generates a sphere of which *OD* is a radius.

Cor. 1. All the radii of a sphere are equal to one another. Therefore,

Def. A sphere is a surface every point on which is equidistant from a fixed point within called the centre.

Cor. 2. The sphere is a closed figure, so that to pass from without the sphere to within, or from within to without, it is necessary to cross the surface.

Cor. 3. A point is within a sphere, on the sphere, or without it, according as its distance from the centre is less than, equal to, or greater than the radius of the sphere.

Cor. 4. Two spheres which have the same centre and the same radius coincide in all their parts and form virtually but one sphere.

77. Theorem. Every plane section of a sphere is a circle.

Let *DEP* be the plane section and *P* be any point on it (Fig. of 76).

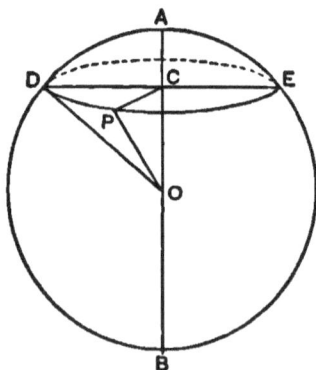

Then, O being the centre of the sphere, OP is constant, and P lies in the plane of section.

Therefore (Art. 10. Cor.) the section is a circle.

Def. The section by a plane through the centre of the sphere is the largest circle producible, and is called a *great circle of the sphere.* All other sections are *small* circles.

Cor. A great circle of a sphere has its centre coincident with that of the sphere; and the generating semicircle of the sphere is one-half of one of its great circles.

78. Theorem. A line can meet a sphere in two, and in only two, points.

Proof. If a line meets a sphere, any plane containing the line gives in section a circle cutting the line; and as the circle cuts the line twice, and twice only, so a line can meet the sphere in two, and in only two, points.

Def. A line which meets a sphere is a *secant* line, and the part within the sphere is a *chord.*

A secant through the centre is a *centre line,* and its chord is a *diameter.*

A plane which cuts a sphere is a *secant plane,* and when it passes through the centre it is a *diametral plane.*

79. Theorem. The join of the centre of a sphere with the middle point of a chord is perpendicular to the chord.

Let DE be a chord whose middle point is C (Fig. of 76); then OC is $\perp DE$.

The plane of O and DE gives in section a circle with DE as chord, and O as centre. And C being the middle point of the chord, OC is $\perp DE$ (P. Art. 96. 4).

Cor. 1. Diameters of the same small circle bisect one another, and being chords of the sphere, the join of the centre of a small circle with the centre of the sphere is normal to the plane of the small circle, *i.e.* to the plane of section.

The converse of this is evidently true.

Cor. 2. Lines through the centres of small circles and respectively normal to their planes meet at the centre of the sphere.

Cor. 3. The plane normal to any chord at its middle point contains the centre of the sphere.

For this plane is the right-bisector plane of the chord, and therefore contains every point equidistant from the end points of the chord. But the centre of the sphere is equidistant from the end points of the chord.

80. Problem. To find the centre of a given sphere.

1st Solution. Draw, on the sphere, two small circles whose planes are not parallel, and draw normals to the planes of these circles at their centres.

These normals meet at the centre of the sphere (Art. 79. Cor. 2).

2d Solution. Draw any three non-parallel chords and their right-bisector planes.

These planes have the centre as their common point (Art. 79. Cor. 3).

Cor. 1. In the first solution, if the planes of the circles are parallel, the normals also are parallel; and as they pass through the same point, the centre of the sphere, they are coincident (P. Art. 70. Ax.).

Therefore the centres of parallel sections of a sphere lie upon a centre line normal to the planes of section, and are therefore collinear.

Cor. 2. In the second solution, if the chords are parallel, so also are their right-bisector planes; and as these planes are concurrent, they are also coincident.

Therefore, the middle points of parallel chords in a sphere are complanar, and lie upon a diametral plane normal to the chords.

81. When the two points in which a line meets a sphere become coincident, the line becomes a tangent line to the sphere and touches the sphere in a double point.

Hence, for a sphere to touch a given line at a given point is equivalent to two conditions.

82. Theorem. A tangent line to a sphere is perpendicular to the radius to the point of contact, and conversely.

Proof. The plane determined by the tangent line and the radius to the point of contact gives in section a circle with its tangent line and radius, and as the same angle is involved, the truth of the theorem follows (P. Art. 110).

Def. An indefinite number of perpendiculars may be drawn to a radius at its extremity; these are all tangent lines, and they all lie in a plane to which the radius is normal.

This plane is a tangent plane to the sphere.

Cor. A tangent plane is normal to the radius to the point of contact.

83. Theorem. Through any four non-complanar points, of which no three are in line, one, and only one, sphere can pass.

Proof. It is shown in Art. 46. Cor. 3, that one, and only one, point is equidistant from four given non-complanar points, no three of which are in line.

If this point be taken as centre, and its distance from any one of the given points be taken as radius, the sphere so determined passes through the four given points.

Cor. 1. Four non-complanar points, no three of which are in line, determine one sphere.

Cor. 2. Spheres which coincide in four non-complanar points coincide altogether.

Def. Four or more points so situated that a sphere can pass through them are *conspheric,* and when these points form the vertices of a figure, the figure is inscribed in the sphere, and the sphere circumscribes the figure.

When a sphere has all the sides of a skew polygon as tangent lines, the sphere is inscribed to the polygon, and the polygon is circumscribed to the sphere.

With a polyhedron it is different. For a sphere may have the edges as tangent lines, or the faces as tangent planes, but not both.

The sphere having the edges as tangent lines is the *tangent sphere to the edges,* and the one having the faces as tangent planes is the *tangent sphere to the faces.*

Cor. 3. A regular polyhedron, on account of its complete symmetry, has all its vertices conspheric, all its edges tangent lines to a sphere, and all its faces tangent planes to a sphere, and these three spheres have the same centre.

84. The vertices of a skew quadrilateral are necessarily conspheric; for from the definition (29) they are four non-complanar points, no three of which are in line.

Let A, B, C, D be the vertices taken in order, and let the sides AB, BC, CD, and DA be considered as lines of indefinite length.

Cor. 1. Let A and B become coincident. Then AB becomes a tangent line to the sphere.

Therefore, a sphere can touch a given line at a given point, and pass through any two other points whose join is not complanar with the given line.

Cor. 2. Let A, B, and C become coincident. Then the lines AB and BC become two tangent lines intersecting on the sphere at B, and these determine a tangent plane.

Therefore, a sphere can touch a plane at a given point and pass through any one other point which does not lie in the plane.

As four points properly situated are necessary to determine a sphere, touching a plane at a given point is equal to three conditions, and the point of contact is thus a triple point.

Cor. 3. Let A and B become coincident at one point, and C and D become coincident at another. Then the line *AB* becomes a tangent line at one point, and the line *CD* a tangent line at another, and these two tangents are not complanar.

Therefore, a sphere may touch two non-complanar lines at any two given points, one in each line.

85. Theorem. The figure of intersection of two spheres is a circle, and the common centre line of the spheres passes through the centre of the circle and is normal to its plane.

Proof. Let *O* and *O'* be the centres of the spheres, and *P* be a point on their figure of intersection *PQR*. Then *OP*, and *O'P*, and *OO'* are constant for all positions of *P*. Therefore, *P* lies on a cone-circle to which *O* and *O'* are vertices, and hence *OO'* passes through the centre, *C*, of the circle, and is normal to its plane.

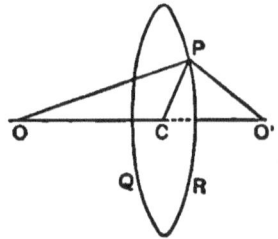

Cor. 1.　*OP* and *O'P* being given, *CP* decreases as *OO'* increases, and *vice versa*. When *OPO'* is a right angle, the tangent planes to the two spheres are perpendicular to one another, and the spheres intersect orthogonally.

Cor. 2.　When *P* comes to *C*, the circle *PQR* becomes a point upon the line *OO'*.

Therefore, when two spheres touch, they do so at a single point, and the common centre line passes through the point of contact.

86. *APBR* is a sphere with *O* as centre and *O'* any point without the sphere. *O'P* is a tangent line from

O', touching the sphere at P. PQR is the small circle through P, whose plane is normal to OO'.

1. OO' and OP are constants, and $\angle OPO'$ is a right angle, since $O'P$ is a tangent (Art. 82). Therefore $O'P$ is constant, and P always lies on the small circle PQR, which is a cone-circle to O and O' as vertices.

Therefore all tangent lines from a given point to a sphere are equal.

2. $O'P$ is the generator of a circular cone which touches the sphere along the small circle PQR, and O' is the centre or vertex of the cone.

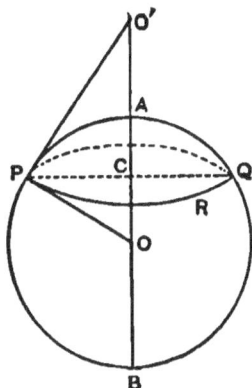

Def. The cone of which $O'P$ is the generator is the *tangent cone* for the point O'.

The circle PQR is the *circle of contact*, and its plane is the *polar plane* of the point O' with respect to the sphere; and the point O' is the *pole* of the plane.

3. When O' comes to A, the tangent cone and the polar plane of O' unite to form the tangent plane at A; hence a tangent plane is a double plane representing the limiting form of the tangent cone, and the limiting position of the polar plane as the pole comes to the sphere.

Evidently, then, a tangent plane is the polar plane to its point of contact.

87. Problem. To find the locus of a point equidistant from three planes, no two of which are parallel, and which do not form an axial pencil; *i.e.* from three planes which form a sheaf.

Let ABC, ACD, ADB be the planes having A as their common point.

Let the internal and external bisecting planes of the dihedral angle whose edge is AB be denoted by $\overline{ab}$ and $\overline{AB}$ respectively, and similarly for the other dihedral angles.

Also, let A_t denote the common line of $\overline{ab}$ and $\overline{ac}$. Then, as every point on $\overline{ab}$ is equidistant from the planes ABC and ABD, and every point on $\overline{ac}$ is equidistant from the planes ACB and ACD, every point on the intersection of $\overline{ab}$ and $\overline{ac}$, that is, on A_t, is equidistant from the three given planes.

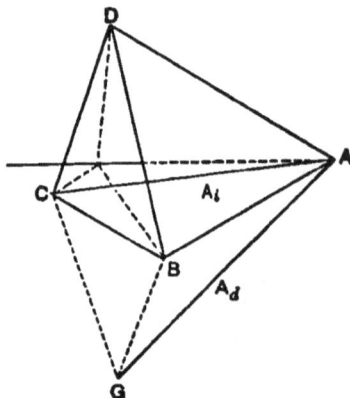

The line A_t is thus inclined to all the planes at the same angle, and it will be called the internal isoclinal line to the planes.

Again, every point on $\overline{AB}$ is equidistant from the planes ABC and ABD, and every point on $\overline{AC}$ is equidistant from the planes ACB and ACD.

Therefore, every point on the common line of $\overline{AB}$ and $\overline{AC}$, that is on the line A_d, is equidistant from the three planes, and A_d is an external isoclinal line to the planes.

Similarly, A_b and A_c are external isoclinals to the same three planes.

Therefore, the required locus consists of the four isoclinal lines to the planes. These isoclinals pass through A, the common point, and form the centre locus of a sphere which touches the three concurrent planes.

Cor. Each isoclinal line is the common line to three bisector planes which form an axial pencil, viz.:

$$A_i \text{ of } \overline{ab}, \overline{ac}, \overline{ad}; \ A_b \text{ of } \overline{ab}, \overline{AC}, \overline{AD};$$

$$A_c \text{ of } \overline{ac}, \overline{AB}, \overline{AD}; \text{ and } A_d \text{ of } \overline{ad}, \overline{AB}, \overline{AC}.$$

88. Problem. To find the centre of a sphere which shall touch four planes so situated as to form a tetrahedron.

Employing the notation of Art. 87, we have four isoclinal lines to three of the planes, at each vertex of the tetrahedron, or 16 in all. These are A_i, B_i, C_i, D_i as internal ones, and $A_b, A_c, A_d, B_a, B_c, B_d, C_a, C_b, C_d, D_a, D_b, D_c,$ as external ones.

Denote the planes opposite A, B, C, D by $\alpha, \beta, \gamma, \delta$, respectively.

Then A_i is the locus of a point equidistant from β, γ, δ; and B_i from α, γ, δ. Therefore, a point equidistant from γ and δ lies upon both A_i and B_i, and hence these lines intersect, and C_i and D_i pass through the point of intersection.

Hence (1) A_i, B_i, C_i, D_i, meet to give one point required.

Similarly, each of the following groups of four lines gives a point equidistant from the four planes:

(2) A_i, B_a, C_a, D_a; (3) B_i, A_b, C_b, D_b;

(4) C_i, A_c, B_c, D_c; and (5) $D_i, A_d, B_d, C_d.$

Again, A_b is the locus of a point equidistant from β, γ, δ, and B_a from α, γ, δ.

Therefore, A_b and B_a intersect in a point equidistant from $\alpha, \beta, \gamma, \delta$, and C_d and D_c pass through this point.

Hence these lines meet in groups of four to give three points equidistant from the four planes ; namely,

(6) A_b, B_a, C_d, D_c ; (7) A_c, C_a, B_d, D_b ; (8) A_d, D_a, B_c, C_b.

Thus eight spheres, in all, can be found, each of which shall touch four planes so situated as to form a tetrahedron.

EXERCISES E.

1. If the director figure in the generation of a cone (61) is a polygon, what figure is formed ?

2. Show that the cone is a limiting case of an n-faced corner, and explain how.

3. If the radius of a sphere is the generator of a circular cone, the figure of intersection of the sphere and cone is a circle.

4. The centre locus of a sphere which touches a plane at a given point is a normal to the plane at the given point.

5. What is the centre locus of a sphere which touches a line at a given point ? which touches two parallel lines ? which touches two intersecting lines ? which touches two intersecting planes ?

PART II.

AREAL RELATIONS INVOLVING LINE–SEGMENTS ABOUT SPATIAL FIGURES.

89. The theorem in plane geometry that the square on the hypothenuse of a right-angled triangle is equal to the sum of the squares on the sides, the theorem that the rectangle on the parts of a secant line between a point and a circle is equal to the square on the tangent from the point to the circle, and others of this nature, express areal relations, involving line-segments of plane figures.

Many important relations of a similar nature exist among the line-segments connected with spatial figures. These we propose to consider in this part of the work.

SECTION 1.

The Skew Quadrilateral and the Polyhedron.

90. Theorem. In a skew quadrilateral the sum of the squares on the sides is greater than the sum of the squares on the diagonals, by four times the square on the join of the middle points of the diagonals.

Proof. $ABCD$ is a skew quadrilateral, and AC and BD are its diagonals, having I and J as their middle points.

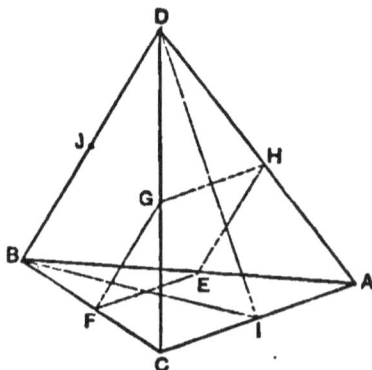

DI is median to $\triangle\,CDA$, and BI is median to $\triangle\,CBA$.

$$\therefore\ CD^2+DA^2+CB^2+BA^2=2\,(CI^2+DI^2+CI^2+BI^2);$$

or, $$\Sigma AB^2 = 4\,CI^2 + 2\,DI^2 + 2\,BI^2.$$

But IJ is median to the $\triangle\,DIB$;

$$\therefore\ 2\,DI^2 + 2\,BI^2 = 4\,BJ^2 + 4\,IJ^2;$$

or, $$\Sigma AB^2 = 4\,CI^2 + 4\,BJ^2 + 4\,IJ^2$$
$$= CA^2 + BD^2 + 4\,IJ^2. \qquad \text{Q. E. D.}$$

This important theorem is true of all quadrilaterals, whether plain or skew (P. Art. 173).

78

91. For the tetrahedron let us adopt the following notation: Taking ABC as the base and D as the vertex, denote the lateral edges DA, DB, DC by a, b, and c respectively, and the basal edges BC, CA, AB by a_1, b_1, and c_1 respectively. Then a and a_1 are opposite edges, etc.

Theorem. In any tetrahedron four times the sum of the squares on the diameters is equal to the sum of the squares on the edges.

Proof. The skew quadrilateral with its diagonals forms the tetrahedron.

The results of Art. 90 give:

$$4\,IJ^2 = a^2 + c^2 + a_1^2 + c_1^2 - b^2 - b_1^2, \text{ (Fig. of 90.)}$$

$$4\,FH^2 = b^2 + c^2 + b_1^2 + c_1^2 - a^2 - a_1^2,$$

$$4\,EG^2 = a^2 + b^2 + a_1^2 + b_1^2 - c^2 - c_1^2.$$

Therefore, by addition,

$$4\,(IJ^2 + FH^2 + EG^2) = a^2 + b^2 + c^2 + a_1^2 + b_1^2 + c_1^2;$$

or, denoting, in general, a diameter by d and an edge by e,

$$4\,\Sigma d^2 = \Sigma e^2.$$

Cor. In the regular tetrahedron, all the diameters being equal, and all the edges being equal, gives,

$$\Sigma d^2 = 3\,d^2, \text{ and } \Sigma e^2 = 6\,e^2;$$

$$\therefore\ e^2 = 2\,d^2.$$

So that if the diameter is equal to the side of a square, the edge is equal to the diagonal of the square (P. Art. 180. Cor.).

92. Theorem. In any tetrahedron, nine times the square on a median is equal to the difference between three times the sum of the squares on the conterminous edges and the sum of the squares on the remaining edges.

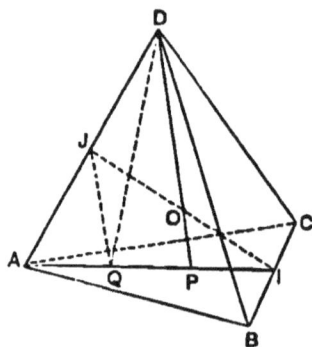

Proof. In the tetrahedron $D \cdot ABC$, P is the centroid of ABC. Then DP is the median to the face ABC.

Bisect AP in Q. And $AQ = QP = PI = \frac{1}{3} AI$.

DQ is median to the $\triangle ADP$;

$$\therefore AD^2 + DP^2 = 2AQ^2 + 2DQ^2. \qquad \text{(P. Art. 173.)}$$

Also, DP is median in the $\triangle QDI$;

$$\therefore DQ^2 + DI^2 = 2QP^2 + 2DP^2.$$

And eliminating DQ^2 between these relations, we obtain

$$3DP^2 = AD^2 + 2DI^2 - \tfrac{2}{3}AI^2.$$

But $\because AI$ is median in the $\triangle ABC$, and DI is median in the $\triangle BCD$,

$$\therefore 2AI^2 = AB^2 + AC^2 - 2BI^2 = c_1^2 + b_1^2 - \tfrac{1}{2}a_1^2,$$

and $\qquad 2DI^2 = DB^2 + DC^2 - 2BI^2 = b^2 + c^2 - \tfrac{1}{2}a_1^2;$

whence $\quad 3DP^2 = a^2 + b^2 + c^2 - \tfrac{1}{3}(a_1^2 + b_1^2 + c_1^2),$

or $\qquad 9DP^2 = 3\Sigma a^2 - \Sigma a_1^2.$ Q. E. D.

93. Theorem. In any tetrahedron, nine times the sum of the squares on the medians is equal to four times the sum of the squares on the edges.

Proof. Let m_1, m_2, m_3, m_4 denote the medians.
Then from 86,

D as vertex, $9 m_1^2 = 3 a^2 + 3 b^2 + 3 c^2 - a_1^2 - b_1^2 - c_1^2$;

A as vertex, $9 m_2^2 = 3 a^2 + 3 b_1^2 + 3 c_1^2 - a_1^2 - b^2 - c^2$;

B as vertex, $9 m_3^2 = 3 a_1^2 + 3 b^2 + 3 c_1^2 - a^2 - b_1^2 - c^2$;

C as vertex, $9 m_4^2 = 3 a_1^2 + 3 b_1^2 + 3 c^2 - a^2 - b^2 - c_1^2$.

Whence by addition, and denoting a median in general by m and an edge in general by e, we have

$$9 \, \Sigma m^2 = 4 \, \Sigma e^2. \qquad \text{Q. E. D.}$$

Cor. 1. In the regular tetrahedron $9 \, \Sigma m^2 = 36 \, m^2$, and $4 \, \Sigma e^2 = 24 \, e^2$;

$$\therefore \; m^2 = \tfrac{2}{3} e^2.$$

Cor. 2. The median in a regular tetrahedron is the same as the perpendicular from the vertex to the base, and denoting it by p, we have

$$p = \tfrac{1}{3} e \sqrt{6}.$$

Cor. 3. Denoting a dihedral angle of the regular tetrahedron by E,

$$\sin E = DP/DI = \tfrac{1}{3} e \sqrt{6} \div \tfrac{1}{2} e \sqrt{3} = \tfrac{2}{3} \sqrt{2}.$$

And $$\cos E = \tfrac{1}{3}.$$

94. In the regular tetrahedron we have the circumscribed sphere, the tangent sphere to the edges, and the inscribed sphere. Denoting the radii of these by R, ρ, and r respectively,

$$R = OD = \tfrac{3}{4} p = \tfrac{1}{4} e \sqrt{6}, \qquad \text{(Art. 52.)}$$
$$\rho = OI = \tfrac{1}{2} d = \tfrac{1}{4} e \sqrt{2}, \qquad \text{(Art. 91. Cor.)}$$
$$r = OP = \tfrac{1}{4} p = \tfrac{1}{12} e \sqrt{6};$$
$$\therefore \; R^2 : \rho^2 : r^2 = 9 : 3 : 1.$$

THE PARALLELEPIPED.

95. Denote the direction edges by a, b, c, and an edge in general by e, and a diagonal in general by d.

Theorem. The sum of the squares on the diagonals is equal to the sum of the squares on the edges.

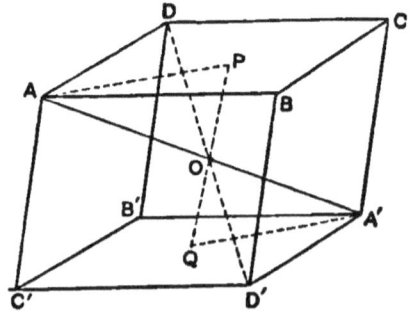

Proof. Since the faces are all parallelograms, and AB is ‖ to $B'A'$, AC to $C'A'$, etc.,

$$AA'^2 + BB'^2 = AB^2 + BA'^2 + A'B'^2 + B'A^2.$$

Similarly,

$$CC'^2 + DD'^2 = CD^2 + DC'^2 + C'D'^2 + D'C^2.$$

Whence, by addition,

$$\Sigma d^2 = AB^2 + CD^2 + A'B'^2 + C'D'^2 + BA'^2$$
$$+ D'C^2 + B'A^2 + DC'^2.$$

And $\quad BA'^2 + CD'^2 = BC^2 + CA'^2 + A'D'^2 + D'B^2;$

and $\quad B'A^2 + C'D^2 = DA^2 + AC'^2 + C'B'^2 + B'D^2.$

$$\therefore \ \Sigma d^2 = \Sigma e^2. \qquad\qquad\qquad \text{Q. E. D.}$$

Cor. 1. As the edges are separable into three groups of four equal edges each (Art. 53),

$$\Sigma d^2 = 4(a^2 + b^2 + c^2).$$

Cor. 2. In the cuboid the diagonals are all equal, and

$$d^2 = a^2 + b^2 + c^2.$$

Cor. 3. In the cube $a = b = c$;

$$\therefore \ d^2 = 3e^2.$$

96. Problem. To find a diagonal of a given parallelc-piped by a plane construction.

In Fig. (1), let $AO'BC$ be the given ppd., and AA' be the diagonal required.

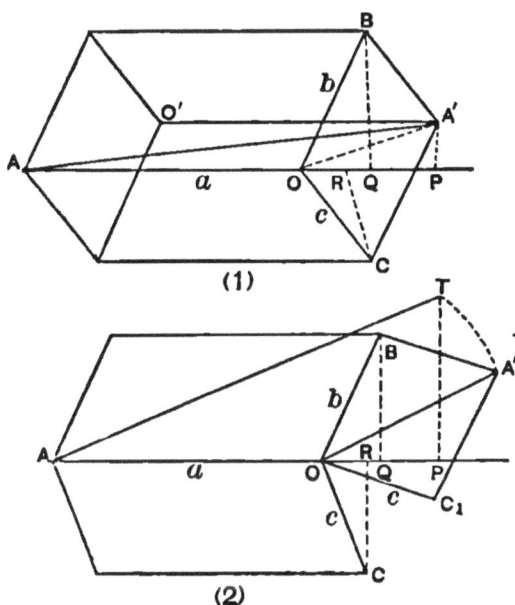

(1)

(2)

Analysis. Draw $A'P$, BQ, CR, perpendiculars on AO, produced if necessary.

P, Q, R are the projections, on AO, of A', B, and C.

The projection of the middle point of OA' is the same point as the projection of the middle of BC, *i.e.* it is the middle point of RQ.

$$\therefore\ OP = OR + OQ.$$

Construction. In Fig. (2) construct the faces AB, AC, and OA', disposed as in the figure.

Draw CR, $BQ \perp$ on AO, produced if necessary.

Take $OP = OR + OQ$, and draw $PT \perp$ to AO.

With O as centre and OA' as radius, describe a circle cutting PT in T. Join AT.

AT is the required diagonal.

Proof. OR is the same for both figures, and so also is OQ, and therefore OP; and AO being the same in both, AP is the same in both.

Also, OT of Fig. (2) is made equal to OA' of (1).

$\therefore \triangle OPT$ of (2) $\equiv \triangle OPA'$ of (1), and $PT = PA'$.

Hence $\triangle APT$ of (2) $\equiv \triangle APA'$ of (1),

and $\qquad AT$ of (2) $= AA'$ of (1).

In like manner any other diagonal can be constructed.

Cor. Let the face angles about the vertex A' be all acute, and the figure is an acute ppd. (Art. 54).

Denote

$\angle BA'C$ by λ, $\angle CA'O'$ by ν, and $\angle O'A'B$ by μ.

Then, Fig. (2),

$\angle BOC_1 = \lambda$, $\angle COP = \mu$, $\angle BOP = \nu$.

Now,

$$AT^2 = AP^2 + PT^2 = (AO + OP)^2 + OA'^2 - OP^2$$
$$= AO^2 + 2\,AO \cdot OP + OA'^2$$
$$= a^2 + 2\,a\,(OQ + OR) + OA'^2.$$

But,

$$OA'^2 = b^2 + c^2 + 2\,bc\cos\lambda, \qquad\qquad \text{(P. Art. 217.)}$$

$$OQ = b\cos\nu,\ OR = c\cos\mu.$$

$$\therefore\ AT^2 = AA'^2 = a^2 + b^2 + c^2 + 2\,bc\cos\lambda + 2\,ca\cos\mu$$
$$+ 2\,ab\cos\nu.$$

And this expresses the square on the longest diagonal, *i.e.* the one extending between the vertices having three acute face angles.

The other diagonals are given by making two angles obtuse in every possible way. They are:

$$a^2 + b^2 + c^2 + 2\,bc\cos\lambda - 2\,ca\cos\mu - 2\,ab\cos\nu,$$

$$a^2 + b^2 + c^2 - 2\,bc\cos\lambda + 2\,ca\cos\mu - 2\,ab\cos\nu,$$

$$a^2 + b^2 + c^2 - 2\,bc\cos\lambda - 2\,ca\cos\mu + 2\,ab\cos\nu.$$

For the diagonals of an obtuse ppd. it is only necessary to change throughout the algebraic sign of every cosine term.

REMARK. In making constructions like the foregoing care must be exercised that every measured segment is taken in its proper sense, or with its proper sign.

By taking the face angles about A' all acute, the perpendiculars $A'P$, BQ, CR, all fall to the right of O. Under a different arrangement of angles, some or all of these might have fallen to the left of O. In any case if M is the middle point of QR, OP is to be taken equal to $2\,OM$, whatever the sign may be.

97. Let $O \cdot ABC$ be a cuboid, and OP be a diagonal. Then $OP^2 = OA^2 + OB^2 + OC^2$ (Art. 95. Cor. 2).

But if OX, OY, OZ, the direction lines of the cuboid, meeting in O, be taken as the three rectangular axes of space (Art. 8. Def. 1), OA is the projection of OP on OX, OB is the projection of OP on OY, and OC, of OP on OZ.

Therefore, the square on any line-segment is equal to the sum of the squares of the projections of the segment on any three mutually perpendicular lines.

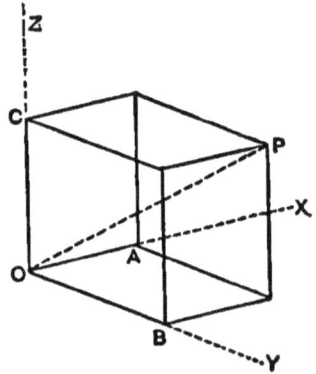

98. Denoting OA by a, OB by b, and OC by c; also $\angle POA$ by α, $\angle POB$ by β, $\angle POC$ by γ, we have

$$\cos^2 \alpha = \frac{OA^2}{OP^2} = \frac{a^2}{a^2 + b^2 + c^2},$$

with the symmetrical expressions for $\cos^2 \beta$ and $\cos^2 \gamma$.

$$\therefore \; \cos^2 \alpha + \cos^2 \beta + \cos^2 \gamma = 1.$$

Def. The angles α, β, γ are *direction angles* of the line OP, and determine the direction of OP relatively to the three axes. The cosines of these angles are the *direction cosines* of OP.

These angles are interdependent, and the result of this theorem shows that the sum of the squares of their cosines is unity.

Cor. The position of a point, P, in space is known relatively to the origin O, and the axes OX, OY, OZ, when we are given OP, and the angles which OP makes with the axes; or when we are given the length of the projections of OP upon the axes. For the projections are the direction edges of a cuboid of which OP is the diagonal.

This is the fundamental principle entering into analytic spatial geometry.

The Octahedron.

99. The octahedron may assume a variety of forms, but we shall confine ourselves to those in which the point of intersection of the axis is the middle point of each axis, or the centre of the figure.

In general the octahedron is the reciprocal of the parallelepiped, formed by joining the centres of adjacent faces.

The three joins of the centres of opposite faces of the ppd. in pairs are the axes of the octahedron, and hence from the nature of a ppd. (Art. 37) the octahedron may be triclinic, diclinic, monoclinic, right or regular; the right octahedron coming from the cuboid, and the regular from the cube.

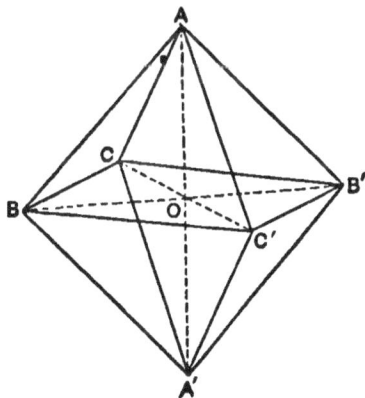

100. Theorem. In any octahedron, the sum of the squares on the twelve edges is equal to twice the sum of the squares on the three diameters AA', BB', and CC'.

Proof. The section along any two diameters, being a parallelogram, gives

$$AA'^2 + BB'^2 = AB^2 + BA'^2 + A'B'^2 + B'A^2,$$

$$BB'^2 + CC'^2 = BC'^2 + C'B'^2 + B'C^2 + CB^2,$$

$$CC'^2 + AA'^2 = AC'^2 + CA'^2 + A'C'^2 + C'A^2.$$

Whence, by addition,

$$2(AA'^2 + BB'^2 + CC'^2) = 2\,\Sigma d^2 = \Sigma e^2.$$

Cor. If the octahedron is regular, all the edges are equal and all the diameters are equal, and therefore

$$d^2 = 2\,e^2,$$

and the section $ACA'C'$ is a square.

The Regular Dodecahedron.

101. Let AE, AB, and AG be the three edges which meet to form the corner of a regular dodecahedron. Let

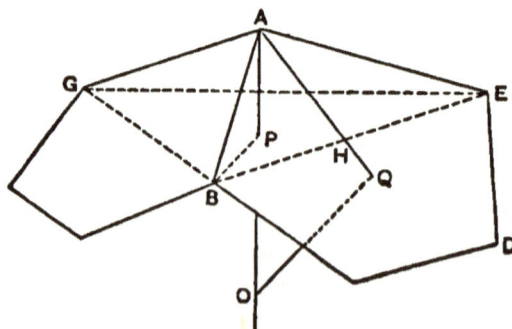

Q be the centre of the face ADB, and O be the centre of the circumscribed sphere.

Since all the faces are congruent, BEG is an equilateral triangle, and OA passes through its centroid P, and is normal to the plane of the triangle (Art. 79. Cor. 1).

$\angle ABE = 36°$, and $BE = 2\,BH = 2\,AB \cos 36° = 2\,e \cos 36°$.

Then, $\qquad BP = \tfrac{1}{3}BE \cdot \sqrt{3} = \tfrac{2}{3}e\sqrt{3}\cos 36°$.

Also, $\because BPA = \urcorner$, $AP^2 = AB^2 - BP^2$,

or $\qquad AP^2 = e^2(1 - \tfrac{4}{3}\cos^2 36°)$.

But if AA' be a diameter of the circumsphere, ABA' is a $\urcorner$, since B is on the sphere.

$\qquad \therefore AA' \cdot AP = AB^2$ (Art. P. 169°);

or $\qquad 2\,R \cdot AP = e^2$.

Whence, $\qquad R = \dfrac{\tfrac{1}{2}e\sqrt{3}}{\sqrt{\{3 - 4\cos^2 36°\}}}.$

Or $\qquad R = \dfrac{e\sqrt{3}}{4\sqrt{\sin 6° \cdot \sin 66°}} = e \times 1.401258\cdots$

Again, we have,

$$\rho = \sqrt{(R^2 - \tfrac{1}{4}AE^2)} = \sqrt{(R^2 - \tfrac{1}{4}e^2)}.$$

$$\therefore \rho = e\sqrt{\{\overline{1.401258}^2 - 0.25\}}$$

$$= e \times 1.309016\cdots,$$

and $\qquad r = \sqrt{\{R^2 - AQ^2\}}.$

But $\qquad AQ = \tfrac{1}{2}AB \sec BAQ$

$$= \tfrac{1}{2}e \cdot \dfrac{1}{\cos 54°} = \tfrac{1}{2} \cdot \dfrac{e}{\sin 36°}.$$

$$\therefore r = e\sqrt{\left\{\overline{1.401258}^2 - \dfrac{1}{4\sin^2 36°}\right\}}$$

$$= e \times 1.113516\cdots.$$

EXERCISES F.

1. Two opposite edges of a tetrahedron are perpendicular to one another when of the remaining edges the sums of the squares upon opposite edges, taken in pairs, are equal.

2. What does the theorem of Art. 91 become when the four vertices of the tetrahedron become complanar?

3. What does the theorem of Art. 92 become when D comes to the centroid of the triangle ABC?

4. Show that the tangent of the angle made by an edge of a regular tetrahedron with one of the faces is $\sqrt{2}$.

5. In the cube, P is the middle point of AB, and S is the middle point of $A'B'$; show that the acute angle of the section through Q, D, S is $\cos^{-1}\frac{1}{3}\sqrt{10}$.

6. In the cube, DK is $\perp$ from D upon the diagonal BB'; show that $DK = \frac{1}{3}e\sqrt{6}$; and that $CK = e$.

7. In the cube, the join of the middle point of AB with B', and the join of the middle point of AD with D', divide each other into parts which are as $2:1$.

8. The angle between two diagonals of a cube is $\cos^{-1}\frac{1}{3}$.

9. In the cube, the angle between a diagonal and a face is $\cos^{-1}\dfrac{1}{\sqrt{3}}$.

10. In the cuboid, the angle subtended at the centre by the middle points of two conterminous edges is

$$\cos^{-1}a^2 / \sqrt{(a^2 + b^2)(a^2 + c^2)},$$

with variations in the letters for the different cases.

11. In the cuboid, the angle between diagonals is

$$\cos^{-1}(a^2 - b^2 - c^2)/(a^2 + b^2 + c^2),$$

with symmetrical variations.

12. In the cuboid, the $\perp$ from a vertex upon a diagonal is

$$a\sqrt{b^2 + c^2} / \sqrt{a^2 + b^2 + c^2},$$

with symmetrical variations.

13. In an octahedron, there may be, at most, six different lengths of edges.

14. If the semi-diameters of an octahedron be a, b, c, and

$$\angle (bc) = \lambda, \quad \angle (ca) = \mu, \quad \text{and} \quad \angle (ab) = \nu,$$

then the squares of the edges are

$$a^2 + b^2 \pm 2\,ab\cos\nu, \quad b^2 + c^2 \pm 2\,bc\cos\lambda, \quad c^2 + a^2 \mp 2\,ca\cos\mu.$$

15. In a right octahedron, the cosines of the dihedral angles are

$$b^2c^2 + c^2a^2 - a^2b^2, \quad c^2a^2 + a^2b^2 - b^2c^2, \quad a^2b^2 + b^2c^2 - c^2a^2,$$

each divided by $\qquad a^2b^2 + b^2c^2 + c^2a^2.$

16. In a regular octahedron, the perpendicular from the centre upon a face is $\frac{1}{6}e\sqrt{6}$.

17. In a regular otcahedron, the cosine of a dihedral angle is $-\frac{1}{3}$.

18. The section through the middle points of AC', AB', $B'C$, CA', $A'B$, and $B'C$ is a hexagon with opposite sides parallel, and is regular if the octahedron is regular.

19. A section of an octahedron parallel to any face is a hexagon.

20. The radius of the tangent sphere to the edges of a regular octahedron is $\frac{1}{2}e$.

21. The squares of the radii of the three spheres of a regular octahedron are in harmonic proportion.

22. In a regular dodecahedron,

$$R = \frac{e}{4}(\sqrt{15} + \sqrt{3}).$$

23. In a regular dodecahedron,

$$r = e \sqrt{\left\{ \frac{25 + 11\sqrt{5}}{40} \right\}};$$

and

$$\rho = \frac{e}{4}(3 + \sqrt{5}).$$

24. In a regular dodecahedron, if D be the dihedral angle,

$$\sin D = \tfrac{2}{5}\sqrt{5}.$$

25. In the regular dodecahedron show that $11\,R^2$ exceeds $15\,r^2$ by $3\,e^2$.

26. In the icosahedron, $R = e\sqrt{\left(\frac{5 + \sqrt{5}}{8}\right)};$

$$\rho = e\sqrt{\left(\frac{3 + \sqrt{5}}{8}\right)}; \text{ and } r = e\sqrt{\left(\frac{7 + 3\sqrt{5}}{24}\right)}.$$

27. In the icosahedron if D be the dihedral angle,

$$\cos D = \tfrac{1}{3}\sqrt{5}, \text{ or } \sin D = \tfrac{2}{3}.$$

28. A sphere touches one face of a regular tetrahedron externally, and the three others internally. Show that its radius is $\tfrac{1}{2}p$; and that the distance from the further vertex at which it touches the three faces is $\frac{e}{3}\sqrt{3}$.

29. If a regular cube and octahedron be circumscribed to the same sphere, their vertices are conspheric.

30. If a regular dodecahedron and icosahedron be circumscribed to the same sphere, their vertices are conspheric.

SECTION 2.

THE SPHERE.

102. *Def.* If P be any point, and a line through P meets a given sphere in A and B, the rectangle $PA \cdot PB$ is called the *power* of the point P with respect to the given sphere.

Cor. A point is without a sphere, on the sphere, or within it, according as the power of the point with respect to the sphere is positive, zero, or negative.

103. The power of a fixed point with respect to a given sphere is independent of the direction of the line whose segments form the rectangle which measures the power.

Proof. Let the line through P meet the sphere in A and B. Since P, A, B are in line, P, A, B, O are complanar, O being the centre of the sphere.

The section by this plane is a great circle, with a secant line through P cutting the circle in A and B. And $PA \cdot PB$ is constant in value for this circle (P. Art. 176). And since all great circles have the same centre and equal radii, $PA \cdot PB$ is constant for every great circle, and therefore for the sphere.

Cor. If A and B become coincident, the secant line becomes a tangent, and the rectangle $PA \cdot PB$ becomes the square on the tangent.

93

Therefore, the power of an external point with respect to a given sphere is the square on the tangent from the point to the sphere; and all tangents from the same point to the same sphere are equal.

104. S and S' are two circles with centres A and B and radical axis L (P. Art. 178).

Let the whole system revolve about the common centre-line AB as an axis, while retaining the fixed relations of the several parts.

The circles describe spheres, and the radical axis, L, describes a plane normal to AB.

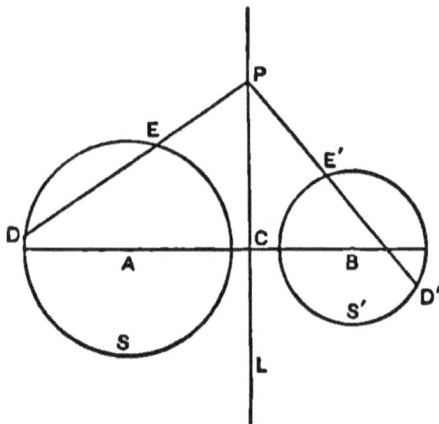

Also $PE \cdot PD = PE' \cdot PD'$ remains true for the spheres. And since P may be any point on the plane described by L, the power of P with respect to each sphere is the same.

Def. The locus of a point of which the power is the same with respect to two given spheres is the *radical plane* of the spheres.

Cor. 1. Evidently, the radical plane of two spheres is normal to the join of their centres, and divides the distance between the centres so that the difference of the squares on the two parts is equal to the difference of the squares on the conterminous radii.

Cor. 2. The tangents to two spheres, from any point on their radical plane, are equal.

Cor. 3. The plane of the circle of intersection of two spheres is their radical plane.

105. Let S_1, S_2, S_3, S_4 be four spheres, and let U_{12} denote the radical plane of S_1 and S_2, etc.

The four spheres have the six radical planes, U_{12}, U_{13}, U_{14}, U_{23}, U_{24}, and U_{34}.

A point whose power with respect to S_1 and S_2 is the same is on the plane U_{12}, and a point whose power with respect to S_1 and S_3 is the same is on the plane U_{13}. Therefore, a point whose power with respect to S_1, S_2, and S_3 is the same is on the common line of U_{12} and U_{13}, and is evidently on the plane U_{23}.

Therefore, the radical planes of three spheres have a common line, and from any point on this line tangents to the spheres are equal.

We shall call this line the *radical line* of the three spheres. In a section through the centres of the spheres, this line gives the radical centre of the three resulting great circles.

Cor. 1. The radical line of three spheres is normal to the plane through their centres.

Cor. 2. The six radical planes to four spheres intersect by threes to form four axial pencils.

The axes of these pencils may be denoted by L_{123}, L_{124}, L_{134}, and L_{234}; L_{123} being the common line to U_{12}, U_{23}, and U_{31}.

Cor. 3. The line L_{123} meets the plane U_{14} in one point only, and it evidently meets U_{24} and U_{34} in the same point.

Therefore, there is, in general, one point from which tangents to four given spheres are equal; or of which the power is the same with respect to four given spheres. This is the *radical centre* of the four spheres.

1. Two secants are drawn through the same point, P, within a sphere, and meet the sphere in A, B, and C, D respectively. Then $PA \cdot PB = PC \cdot PD$.

2. If a, b be the parts into which the plane of a small circle divides the diameter through its centre, the area of the small circle is πab.

3. If three spheres intersect two and two, the planes of the small circles of intersection form an axial pencil.

4. If four spheres intersect two and two, the planes of the circles of intersection pass through a common point.

5. Where is the radical centre of four spheres whose centres are complanar?

6. Under what condition will four spheres have a line of radical centres?
(The spheres are then coaxal.)

7. The tangent cones, common to three spheres taken two and two, have their vertices collinear.

8. The tangent cones, common to four spheres taken two and two, have their vertices complanar.

9. If P and Q be two points in the line L, and U and V intersecting in M be the polar planes of P and Q with respect to a sphere, then every plane through M is polar to some point in L; and L and M are perpendicular to each other.

10. Any rectilinear figure has a corresponding rectilinear figure such that every side in the first figure has a side perpendicular to it in the second.

PART III.

106. A closed spatial figure includes within its boun-
daries a portion of space separated from all other parts of
space. This portion of space considered with respect to
extent, and not with respect to form, is called the *volume*
of the closed figure.

As our primary ideas of a spatial figure were probably
derived from concrete objects such as blocks of wood or
stone, the volume of a spatial figure is also called its
solid contents, and the figure itself is called a *solid.* Hence
the name Solid Geometry.

Also considering a closed spatial figure as a surface,
which after the manner of a closed vessel might be filled
with a liquid, the volume is sometimes called the *capacity*
of the figure.

The measuring of volumes, or solid contents, or capaci-
ties is called *Stereometry.*

Def. Equal spatial figures are those which have equal
volumes, and therefore congruent figures, when having
volumes, are necessarily equal.

97

SECTION 1.

POLYHEDRA.

107. Theorem. Two cuboids with congruent bases have their volumes proportional to their altitudes.

$B \cdot ACD$ and $F \cdot EGH$ are two cuboids having their bases congruent. Then vol. $B \cdot ACD$: vol. $F \cdot EGH$ $= BD : FH$.

Proof. If this proportion is not true, let

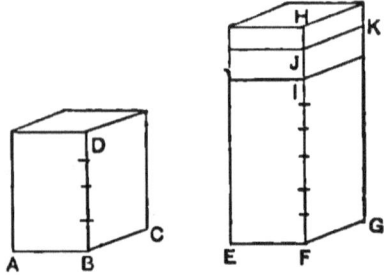

$$\text{vol. } B \cdot ACD : \text{vol. } F \cdot EGH = BD : FI,$$

where FI is different in length from FH; and first let FI be less than FH. As a general case let BD and FH be incommensurable.

Take some *u.l.* (P. Art. 150. 3) less than IH which will measure BD, and divide BD and FH into parts equal to this *u.l.* One point of division, at least, must fall at some point, J, between I and H.

Through all the points of division pass planes parallel to the bases. These divide the cuboids $B \cdot ACD$ and $F \cdot EGJ$ into congruent and therefore equal cuboids.

$$\therefore \text{ vol. } B \cdot ACD : \text{vol. } F \cdot EGJ = BD : FJ$$

and vol. $B \cdot ACD$: vol. $F \cdot EGH = BD : FI$ (hyp.).

$$\therefore \text{ vol. } F \cdot EGJ : \text{vol. } F \cdot EGH = FJ : FI.$$

98

But vol. $F \cdot EGJ <$ vol. $F \cdot EGH$;

$$\therefore FJ \text{ is} < FI;$$

which is not true.

Hence FI cannot be less than FH. And in like manner it is shown that FI cannot be greater than FH; and as FI has some value, it must be equal to FH; and therefore

$$\text{vol. } B \cdot ACD : \text{vol. } F \cdot EGH = BD : FH.$$

Cor. Cuboids which have two dimensions in each respectively equal have their volumes proportional to their third dimensions; or, more generally, a cuboid with constant base has its volume varying as its altitude.

108. Theorem. Two cuboids are to one another as the continued product of their three dimensions.

Let X, Y denote two cuboids whose dimensions are respectively abc, and $a'b'c'$.

Then $X : Y = abc : a'b'c'$.

Proof. Let P be a cuboid whose dimensions are a, b, c', and Q be a cuboid whose dimensions are a, b', c'.

Then X and P have the face ab the same, and P and Q have the face ac' the same, and Q and Y have the face $b'c'$ the same.

$$\therefore X : P = c : c', \qquad \text{(Art. 107. Cor. 1.)}$$
$$P : Q = b : b',$$
$$Q : Y = a : a'.$$

Whence, by compounding the three proportions,

$$X : Y = abc : a'b'c'.$$

Cor. 1. A generalised statement of the theorem is, the volume of a cuboid varies as the continued product of its three dimensions.

Cor. 2. When the cuboids are similar, their homologous edges are proportional, and if a', b', c' be homologous to a, b, c,

$$\frac{abc}{a'b'c'} = \frac{a^3}{a'^3} = \frac{b^3}{b'^3} = \frac{c^3}{c'^3}.$$

Therefore, two similar cuboids are to one another as the cubes upon two homologous line-segments.

Or, the volume of a cuboid of constant form varies as that of the cube on any one of its line-segments.

109. In measuring volumes we take as a unit the volume of the cube whose edge is the unit-length. This volume is the *unit-volume,* and it will be denoted by *u.v.*

The three units of extension are thus interconnected, so that the giving of any one of them gives all.

Thus if a cube has its edge taken as the *u.l.*, the area of one of its faces is the *u.a.*, and its volume is the *u.v.*

If the edge of a cube be n unit-lengths, each of its faces contains n^2 unit-areas, and its volume contains n^3 unit-volumes (comp. P. Art. 151).

110. Theorem. The number of *u.v.*s in a cuboid is the continued product of the numbers of *u.l.*s in its three direction edges, or its three dimensions.

Proof. Let X, Y be the cuboids having their three direction edges expressed by a, b, c and a', b', c'.

Then $\qquad X : Y = abc : a'b'c'.$ \hfill (Art. 108.)

Now let $a' = b' = c' =$ one $u.l.$ Then Y contains one $u.v.$ And hence :

The number of $u.v.$s in $X =$ the number of $u.l.$s in $a \times$ the number of $u.l.$s in $b \times$ the number of $u.l.$s in c.

This result is generally expressed by saying that the volume of a cuboid is the product of its three dimensions, an expression of which the full meaning is given above.

Cor. If a, b, c be the three direction edges of a cuboid, ab denotes the area of the face whose edges are a and b, and c is the altitude to that face taken as base.

Therefore, the volume of a cuboid is the product of the area of its base multiplied by its altitude.

111. The product form of three quantitative symbols, where the symbols denote line-segments, is to be interpreted as the volume of the cuboid having for its three direction edges the line-segments denoted by these symbols.

Hence such expressions as abc, $(a + b)ab$, etc., denote volumes of cuboids in geometry, and are consequently said to be of three dimensions even in algebra.

This exhausts the geometry of space as we know it, for space has, for us at least, only three dimensions.

112. Expressions such as $abcd$, or a^2b^2, or a^2bc, etc., are, in algebra, said to be of four dimensions; but when the letters are line-symbols, no interpretation is possible in real geometry.

They may be then said to belong to a hypothetical or imaginary something, which to us can have no real existence, but which is spoken of as geometry of four dimensions, or as space of four dimensions.

In using the symbols and forms of algebra to deduce geometric relations, expressions of four or of higher dimensions may occur as intermediate steps in some transformation, but never as final results.

The associative law in algebra which tells us that $a \cdot bc$, $ab \cdot c$, $b \cdot ac$ are equal, tells us in geometry that the measure of the volume of a cuboid is independent of which face is taken as a base.

The expression a^2b is the cuboid having the square on side a as base and b as altitude, or the cuboid having the rectangle ab as base and a as altitude; and these are the same cuboid differently viewed.

The forms $\sqrt{abc}$, $\sqrt{a^2b}$, etc., are not geometrically interpretable; but $\sqrt{abcd}$ is an area.

The form $\sqrt[3]{abc}$ denotes a line-segment, the edge of the cube whose volume is equal to the cuboid whose dimensions are a, b, c.

PARALLELEPIPED.

113. Theorem. A parallelepiped is equal to the cuboid which has its base and altitude respectively equal to those of the parallelepiped.

We prove this theorem by showing that any parallelepiped can be transformed, without change of volume, into a cuboid having a base and altitude equal to those of the ppd. Let $A \cdot A'BD$ be a triclinic ppd.

Cut it by a plane, EFG, normal to the direction edge AA'.

This section is a parallelogram, and BFE, CGF, FEA', etc., are ⌐s.

On AA' produced take $A'E'=AE$, and through E' pass the plane $E'F'G'$ ∥ to EFG.

Then the corners $A \cdot BDE$ and $A' \cdot B'D'E'$ are evidently congruent, since they are composed of equal face angles disposed in the same order. And if the figure $A \cdot BGE$ be so placed that A coincides with A', AD with $A'D'$, and AB with $A'B'$, this figure will coincide completely with $A' \cdot B'G'E'$, and the ppd. AC' is transformed to the monoclinic ppd. EG', without change of volume, and the base EH' is equal to the base AD', and the altitude remains unchanged.

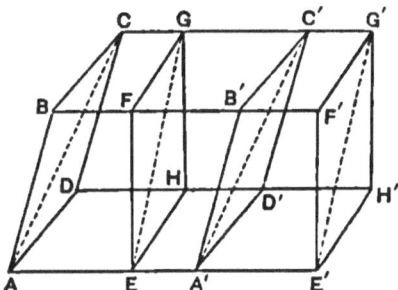

Again, by passing a plane normal to the direction edge EH of the monoclinic ppd. we transform it into a cuboid in which the volume is unchanged, and the base and altitude are unchanged.

Therefore any ppd. can be transformed, without change of volume, into a cuboid having its base and altitude equal to those of the ppd.

Therefore a ppd. is equal to the cuboid having its base and altitude equal to those of the ppd.

NOTE. — If the ppd. is such that it is impossible to cut it by the plane EFG, normal to AA', then EFG may be any plane less inclined to AA' than the face ABC is. We thus transform the ppd. into another triclinic ppd. less oblique than the original; a second section may now be made normal to a direction edge; or if not a second, a third, etc.

Cor. 1. The volume of a parallelepiped is the product of the area of its base by its altitude.

Cor. 2. Similar ppds. are to each other as the cubes on homologous line-segments.

Cor. 3. A ppd. of constant form varies as the cube on any of its line-segments.

PRISM.

114. If a cuboid or a monoclinic ppd. be divided into two triangular prisms by a plane passing through a pair of opposite edges, which are normal to a face, the prisms so formed are congruent, and therefore equal. But if a plane be passed through opposite edges of a triclinic ppd., the two prisms formed are, in general, not congruent, but symmetrical, and they cannot therefore be shown to be equal by superposition. We proceed to show that they are, however, equal.

115. Theorem. The two triangular prisms into which a parallelepiped is divided by a plane through a pair of opposite edges, are equal.

Let $A \cdot BDA'$ be a tri-clinic ppd. A, A', C', C are complanar, and their plane divides the ppd. into the triangular prisms $A \cdot BCA'$ and $C \cdot ADC'$.

These prisms are not congruent. But, as in Art. 113, transforming the triclinic ppd. into the monoclinic ppd. $E \cdot FHE'$, we transform, without change of volume,

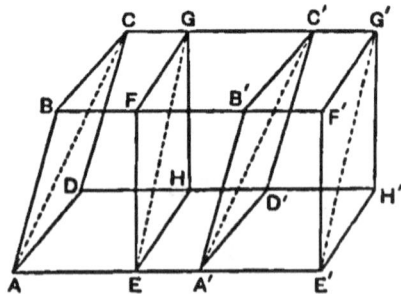

the prism $A \cdot BCA'$ into $E \cdot FGE'$, and $C \cdot ADC''$ into $G \cdot EHG'$. And these new prisms, being right prisms from the monoclinic ppd., are congruent, and therefore equal, and each is one-half the monoclinic ppd.

Therefore, the original prisms $A \cdot BCA'$ and $E \cdot FGE'$ are equal, and each is one-half the triclinic ppd.

Cor. 1. Since the right section $EFGH$ is double the right section EGH, it follows that the volume of a prism is the area of a right section multiplied by the length of a lateral edge.

Cor. 2. Taking $ABCD$ as the base of the ppd., and ACD as the base of the prism $C \cdot ADC''$, these figures have the same altitude.

Therefore (Art. 113. Cor. 1), the volume of a prism is the area of the base multiplied by the altitude.

Cor. 3. As all prisms may be divided into triangular prisms, Cors. 1 and 2 are true for all prisms.

116. We have two expressions for the volume of a prism:

1st,　　vol. = area of rt. section × lateral edge.

2d,　　vol. = area of the base　× altitude.

$$\therefore \frac{\text{area of rt. section}}{\text{area of base}} = \frac{\text{altitude}}{\text{lateral edge}}.$$

But if AA' be the lateral edge, and AP be the altitude, $AP \div AA'$ is the cosine of the angle between the lateral edge and the altitude.

But, as the altitude is normal to the base, and the lateral edge is normal to a right section, this is the angle between a right section and the plane of the base. Calling this the *angle of obliquity* of the prism, we have:

The area of a right section of a prism is equal to the area of an oblique section multiplied by the cosine of the angle of obliquity.

OF LAMINÆ.

117. To fix our ideas, let *AETN* be a trapezoid with *ET* ‖ to *AN*. Divide its side *AE* into any number of equal parts, and through the points of division, *B, C, D*, etc., draw lines ‖ to *AN*.

On these lines and the bases *AN* and *ET* construct the series of internal rectangles *BP, CQ, DR, ES,* ⋯ and the series of external rectangles *Aa, Bb, Cc, Dd* ⋯.

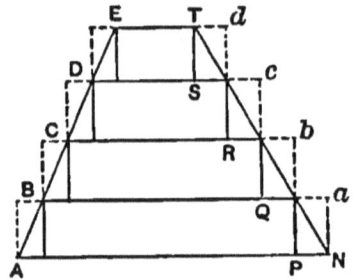

The area of the trapezoid evidently lies between the sum of the external rectangles and the sum of the internal rectangles.

Now, any external rectangle as *Cc* is congruent with an internal rectangle below it, *CQ*; except that the lowest external rectangle has no corresponding and congruent internal one, and the uppermost internal rectangle has no congruent external one.

Let *E* denote the sum of the external rectangles, and *I* denote the sum of the internal ones. Then

$$E - I = \square Aa - \square ES.$$

∴ the difference between the sum of the external rectangles and the sum of the internal rectangles is less than the lowermost external rectangle; and this is true however many rectangles be formed.

But the lowermost rectangle can be made as small as we please, by making its altitude sufficiently small; *i.e.* by making the number of parts into which we divide AE sufficiently great. And hence the area of the trapezoid is the *limit* of the sum of either series of rectangles as the number of rectangles is indefinitely increased.

118. Now, let $AETN$ be a vertical section of a frustum of a pyramid (Art. 59), in which AN and ET are sections of the bases. Divide AE into any number of equal parts, and through the points of division pass planes parallel to the bases.

On the figures of section construct a series of inscribed prisms, BP, CQ, DR, $ES \cdots$, and a series of circumscribed prisms, Aa, Bb, Cc, $Dd \cdots$.

The volume of the frustum lies between the sum of the internal prisms and the sum of the external prisms.

But any external prism, except the lowermost, has a congruent internal prism below it, and any internal prism, except the uppermost, has a congruent external prism above it.

Hence if E denotes the sum of the external prisms, and I of the internal prisms,

$$E - I = \text{prism } Aa - \text{prism } ES$$

$$= \text{vol. of lowermost external prism}$$

$$- \text{vol. of the uppermost internal prism.}$$

And this is true, however many equal parts AE is divided into.

Therefore, the volume of the frustum differs from the sum of either series of prisms, by less than the volumes of the series of prisms differ from each other; that is, by a quantity less than the lowermost external prism; and this difference may be made as small as we please by dividing AE into a sufficiently large number of parts.

Hence, the volume of the frustum is the *limit* of that of either series of prisms, when the number of prisms is indefinitely increased.

Cor. This theorem is exceedingly important, for the least consideration will show that nothing in the investigation requires that AE, or any edge, should be a straight line, and hence that the theorem holds true when the boundary of the figure, between the parallel bases, is composed partly or wholly of curved surfaces; also that the theorem is true when one or both bases reduce to lines or points.

119. *Def.* When a spatial figure is cut by two indefinitely near parallel planes, the prism, having one of the sections as base, and the distance between the planes as altitude, is called a *lamina* of the spatial figure.

When two figures are confined between the same two parallel planes, the laminæ determined by two indefinitely near planes, parallel to the confining planes, are *corresponding laminæ.*

Usually the planes which determine a lamina are supposed to be infinitely near, so that a lamina is one of the prisms of the preceding article, taken at its limit.

Cor. 1. From Art. 118, it appears that two figures which have all corresponding laminæ equal are them-

selves equal; and two figures which have all corresponding laminæ in the same ratio are themselves in
that ratio, the one to the other.

Cor. 2. Since corresponding laminæ have the same
altitude, their volumes are proportional to their bases;
and hence corresponding laminæ are equal when corresponding sections are equal; and corresponding laminæ
are in the same proportion to one another as are the corresponding sections.

THE PYRAMID.

120. Theorem. Pyramids are equal whose bases are
equal and whose altitudes are equal.

Proof. Let the trian-
gular pyramids, $D \cdot ABC$
and $H \cdot EFG$, have their
bases equal, and also their
altitudes equal, and let
them be so placed that
their bases are compla-
nar, and their vertices are
upon the same side of this
plane. Then D and H lie
in a plane parallel to the
plane of the bases.

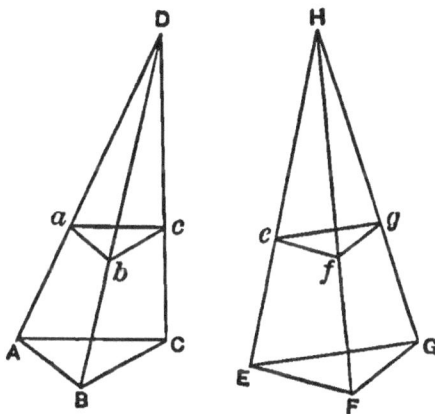

Let abc and efg be corresponding sections.

Then (Art. 28. Cor. 2)

$$\triangle abc \backsim \triangle ABC, \text{ and } \triangle efg \backsim \triangle EFG.$$

But (P. Art. 218. 2), $\triangle abc : \triangle ABC = ab^2 : AB^2$.

And since DA and DB are cut by parallel planes, AB is ∥ ab.

$$\therefore\ ab^2 : AB^2 = Da^2 : DA^2.$$

And   (Art. 27) $Da : DA = He : HE$;

$$\therefore\ ab^2 : AB^2 = Da^2 : DA^2 = He^2 : HE^2 = ef^2 : EF^2,$$

or $\triangle\, abc : \triangle\, ABC = \triangle\, efg : \triangle\, EFG.$

But $\triangle\, ABC = \triangle\, EFG$;   (hyp.)

$$\therefore\ \triangle\, abc = \triangle\, efg.$$

And as corresponding laminæ are equal, the volumes of the pyramids are equal (Art. 119. Cor. 1).

And since all pyramids may be divided into triangular pyramids,

Therefore, any two pyramids are equal whose bases are equal and whose altitudes are equal.

Cor. Two frustums of pyramids which have their two bases respectively equal and their altitudes equal are themselves equal.

121. Theorem. A triangular prism can be divided into three equal pyramids.

Proof. $A \cdot BCD$ is a triangular prism. Pass a plane through the points A, C, and E, and another plane through C, D, and E.

The prism is divided into three equal pyramids.

For $C \cdot FDE$ and $E \cdot CAB$ have their bases DEF and ABC equal, and their altitudes the same as that of the prism. These pyramids are therefore equal (Art. 120).

Also the pyramids $C \cdot ADE$ and $C \cdot ABE$ have their bases ADE and ABE equal (P. Art. 141. Cor. 1), and have their vertices coincident. Therefore they have the same altitude and are equal.

And the prism is thus divided into three equal pyramids.

Cor. 1. As each triangular pyramid is one-third of the corresponding triangular prism, and as every prism can be divided into triangular prisms;

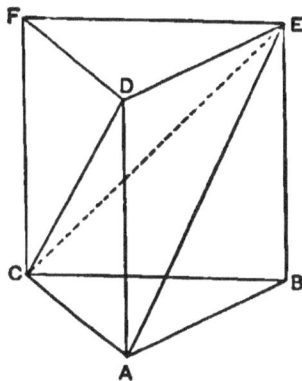

Therefore every pyramid is one-third of the prism having the same base and altitude as the pyramid.

Cor. 2. If B denotes the area of the base of a pyramid, and h denotes its altitude,

$$\text{vol. of pyramid} = \tfrac{1}{3}\, hB.$$

Cor. 3. Pyramids with equal bases are to one another as their altitudes, and pyramids with equal altitudes are as their bases.

122. Theorem. The frustum of a triangular pyramid may be divided into three pyramids, two of which have the bases of the frustum as their bases, and the altitude of the frustum as their altitude, and the third of which is a mean proportional between the first two.

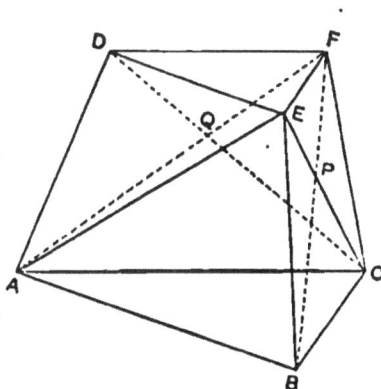

ABCDEF is a triangular frustum.

The plane through A, E, F cuts off the pyramid $A \cdot DEF$, whose base DEF is the upper base of the frustum.

From the remaining figure the plane through A, E, C cuts off the pyramid $E \cdot ABC$, whose base ABC is the lower base of the frustum.

We have left the pyramid $E \cdot AFC$.

Join BF and CD.

The pyramids $B \cdot AEC$ and $F \cdot AEC$ having the common base AEC are as their altitudes, and the altitudes are as PB to PF, or BC to EF.

$$\therefore\; B \cdot AEC : F \cdot AEC = BC : EF.$$

Again, the pyramids $C \cdot AEF$ (which is the same as $F \cdot AEC$) and $D \cdot AEF$ having the common base AEF are to one another as CQ is to QD, or AC to DF.

But, since the bases are similar (Art. 28. Cor. 2),

$$BC : EF = AC : DF.$$

$$\therefore\; B \cdot AEC : F \cdot AEC = F \cdot AEC : D \cdot AEF.$$

Or the pyramid $F \cdot AEC$, or $C \cdot AEF$, is a mean proportional between the pyramids $E \cdot ABC$ and $A \cdot DEF$.

Cor. If B and B' denote the bases of the frustum, and h the altitude,

$$\text{vol. of } E \cdot ABC = \tfrac{1}{3}hB, \text{ vol. of } A \cdot DEF = \tfrac{1}{3}hB',$$

and $\therefore$ vol. of $F \cdot AEC = \tfrac{1}{3}h\sqrt{BB'}$.

The volume of the frustum is accordingly :

$$\text{vol.} = \tfrac{1}{3}h\{B + B' + \sqrt{BB'}\}.$$

123. The volume of the frustum may also be found as follows :

Let O be the vertex of the pyramid from which the

frustum is formed, and let OP be the altitude of the pyramid. Also let OP' be the altitude of the pyramid $O \cdot DEF$, which is removed in forming the frustum. Then,

The frustum $=$ pyr. $O \cdot ABC -$ pyr. $O \cdot DEF$,

and $$OP - OP' = h.$$

Since any area may be expressed as a square, let $b^2 = B$ or the base ABC, and $b'^2 = B'$ or the base DEF.

Then $$OP : OP' = OA : OD = AB : DE = b : b',$$

and $$OP - OP' : OP = b - b' : b.$$

$$\therefore \ OP(b - b') = bh.$$

$$\therefore \ OP = \frac{bh}{b - b'}, \text{ and } OP' = \frac{b'h}{b - b'};$$

and

$$\text{frust.} = \tfrac{1}{3} OP \cdot b^2 - \tfrac{1}{3} OP' \cdot b'^2$$

$$= \tfrac{1}{3} h \left(\frac{b^3 - b'^3}{b - b'} \right) = \tfrac{1}{3} h (b^2 + b'^2 + bb')$$

$$= \tfrac{1}{3} h (B + B' + \sqrt{BB'}).$$

Cor. The volume of a frustum of a pyramid is the sum of the two bases and a mean proportional between the bases, multiplied by one-third of the altitude.

124. *Def.* A triangular prism with non-parallel bases is called a *truncated triangular prism*, or a *wedge.*

Let ABC, DEF be the bases of the wedge, of which ABC is normal to the lateral edges.

Through D pass a plane $\parallel$ to ABC and draw $AP \perp$ to BC.

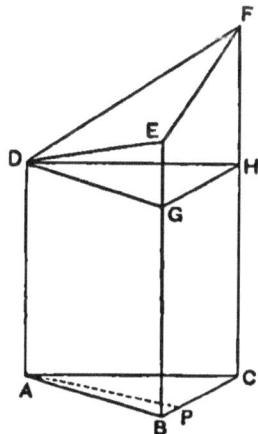

The wedge = the prism $A \cdot BCD$ + the pyramid $D \cdot EFGH$. But the prism $= \triangle ABC \times AD$; and the pyramid

$$= \tfrac{1}{3} AP \times \text{trapezoid } EH,$$

$$= \tfrac{1}{3} \triangle ABC (BE + CF - 2AD);$$

$\therefore$ vol. of wedge $= \tfrac{1}{3} \triangle ABC (AD + BE + CF).$

Or if e_1, e_2, e_3 denote the edges, and B the area of a right section,

$$\text{vol.} = \tfrac{1}{3} B (e_1 + e_2 + e_3).$$

125. Theorem. If a tetrahedron be cut by a plane which bisects two edges and passes through an opposite vertex, the volume of the tetrahedron is equal to four-thirds of the prism having the section as base, and the perpendicular from any other vertex on the plane of section as altitude.

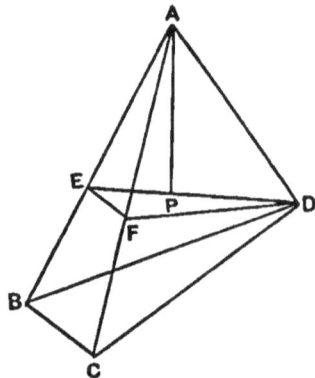

$A \cdot BCD$ is a tetrahedron, and E and F are middle points of AB and AC respectively.

AP is perpendicular upon the plane EFD.

Then $\because EF$ is $\parallel$ to BC, and bisects AB, EF is one-half BC, and $\triangle AEF = \tfrac{1}{4} \triangle ABC$ (P. Art. 218. 2).

The pyramids having these triangles as bases have D as a common vertex;

$\therefore$ tetrahedron $A \cdot BCD = 4$ tetr. $A \cdot DEF$

$$= \tfrac{4}{3} AP \cdot \triangle DEF. \qquad \text{Q. E. D.}$$

PRISMATOID AND ALLIED FORMS.

126. *Def.* A polyhedron with two parallel polygonal bases, and all its lateral faces plane rectilinear figures, and all its lateral edges the joins of vertices of opposite bases, is a *prismatoid.*

This definition includes the prism, pyramid, and frustum of a pyramid as special cases, and is more general than any of these.

When none of the faces are triangles, the figure is the frustum of a pyramid, or a *prismoid,* according as the lateral edges are, or are not, concurrent when produced.

127. *ABCD* and *EFG* are parallel bases of a prismatoid, and *AEB, EBF, FBC, CFG,* etc., are triangular faces, which, in the figure given, are seven in number.

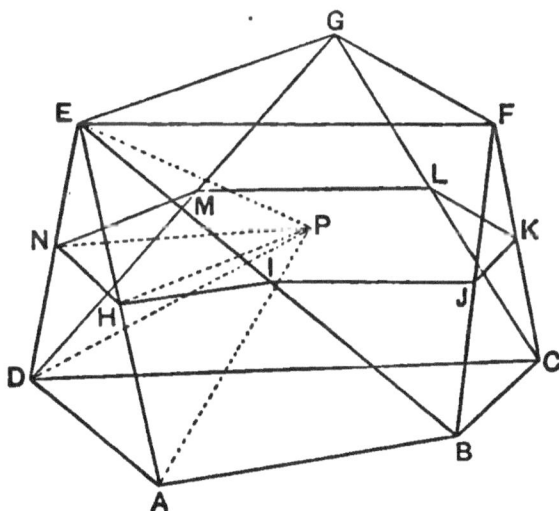

If *n* denotes the number of sides in one base, and *n'* in the other, it is readily seen that the number of faces cannot be greater than $n + n'$.

But if an edge of one base be connected by lateral edges with a parallel edge of the other base, two triangular faces become a quadrangular face, and the whole number of faces is reduced by one. Thus, if EF were parallel to AB, the edges AE, EB, and BF would be complanar, and the two triangular faces AEB and BEF would become one quadrangular face, $AEFB$.

If two other edges of the bases become parallel, a like reduction may take place, and the whole number of faces be reduced by two.

And finally, if the bases have the same number of sides, and each edge in one base be connected with a parallel edge in the other, all the faces become quadrangular, and the figure becomes a frustum of a pyramid or a prismoid, according as the edges, when produced, are or are not concurrent.

Even with the same bases, however, the general appearance of the figure will vary with the different ways of connecting the vertices of the bases by the lateral edges.

128. *Def.* Take H, I, J, etc., middle points of the lateral edges, AE, BE, BF, etc., respectively.

Since HI is parallel to AB (P. Art. 84. Cor. 2. 2) and IJ is parallel to EF, and JK to BC, etc., it follows that H, I, J, etc., lie in a plane which bisects all the lateral edges, and is parallel to the bases. The section by this plane is called the *middle section*.

The middle section contains, at most, $n + n'$ sides, there being always as many sides as there are faces in the prismatoid.

The middle section may contain re-entrant angles, although no such angles are found in either base; and it will frequently have such angles when the bases are polygons of different species, or when their vertices are connected in some particular order.

Cor.   The middle section bisects the altitude.

129. *Volume of the prismatoid.* Take *P*, any point in the plane of the middle section, and join it to *A*, *D*, *E*, *H*, and *N*. Denote the altitude of the prismatoid by *h*.

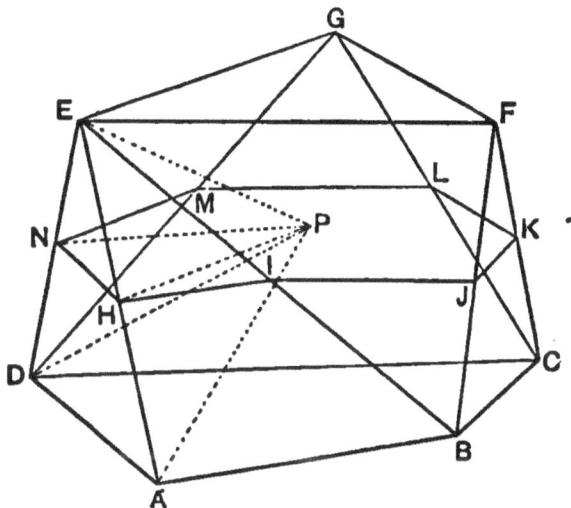

Then, *P·ADE* is a tetrahedron, and *PNH* is a section through a vertex, *P*, and the middle points, *H* and *N*, of two opposite edges.

$$\therefore \text{ vol. of } P\cdot ADE = \tfrac{2}{3}h \times \triangle PNH. \quad \text{(Art. 125.)}$$

Similarly, by joining *P* to all the remaining vertices, *B*, *C*, *F*, etc., and to the remaining middle points, *I*, *J*, *K*, etc., we have,

Sum of all the tetrahedra of which *P·ADE* is the type $= \tfrac{2}{3}h \times$ (the sum of the $\triangle$ of which *PNH* is the type), or

$$\Sigma(P\cdot ADE) = \tfrac{2}{3}h\Sigma(\triangle PNH).$$

But $\Sigma\,(\triangle\,PNH) =$ the area of the middle section, and denoting the area of the middle section by M,

$$\Sigma\,(P \cdot ADE) = \tfrac{2}{3}\,hM.$$

Now, after removing all these tetrahedra, we have left two pyramids having P as a common vertex, and the bases of the prismatoid as their respective bases. The altitude of these prisms being $\tfrac{1}{2}\,h$ (Art. 128. Cor.), their volumes are $\tfrac{1}{6}\,hB$ and $\tfrac{1}{6}\,hB'$, where B and B' are the areas of the bases of the prismatoid.

$$\therefore \text{ vol. of prismatoid} = \frac{h}{6}\,(B + B' + 4\,M).$$

Cor. The prismatoid is equal to four pyramids, two having the bases of the prismatoid as their bases and half the altitude of the prismatoid as their altitude, and two having the middle section as their bases and the altitude of the prismatoid as their altitude.

Cor. The formula of the present article is known as the *prismoidal formula.* On account of its extremely wide range of applicability it is the most important of all formulæ connected with the determination of the volumes of the more prominent spatial figures.

The following examples are some illustrations of its application.

(a) *Prism.* Here the two bases and the middle section are all congruent.

Hence, vol. $= \dfrac{h}{6}\,(B + B + 4\,B) = hB.$ (Art. 115. Cor. 2.)

(b) *Pyramid.* The upper base vanishes, and the middle section is one-fourth the lower base.

$$\therefore \text{ vol.} = \frac{h}{6}(B + 0 + B) = \tfrac{1}{3}hB. \qquad \text{(Art. 121. Cor. 2.)}$$

(c) *Frustum of a pyramid.*

Let B, B' be the bases, and M be the middle section. And since any area may be expressed as a square, let $B = b^2$, $B' = b'^2$, and $M = m^2$.

Then $\qquad 2\,m = b + b'.$

$$\therefore\ 4\,m^2 = 4\,M = b^2 + b'^2 + 2\,bb'$$

$$= B + B' + 2\sqrt{BB'}.$$

$$\therefore\ \frac{h}{6}(B + B' + 4\,M) = \frac{h}{6}(2\,B + 2\,B' + 2\sqrt{BB'})$$

$$= \frac{h}{3}(B + B' + \sqrt{BB'}). \quad \text{(Art. 122. Cor.)}$$

(d) *Tetrahedron,* in terms of a middle section (Art. 51. Def. 1) and the length of the common perpendicular to the edges parallel to the section.

In this case, which has an important subsequent application, both bases vanish, and we have

$$\text{vol.} = \tfrac{2}{3}\,hM.$$

REGULAR POLYHEDRA.

130. A regular polyhedron of n faces is divisible into n congruent pyramids whose bases are the several faces of the polyhedron, and whose altitude is the radius of the in-sphere to the polyhedron.

Hence, if n be the number of faces, B be the area of a face, and r be the radius of the in-sphere, we have

$$\text{vol.} = \frac{n}{3} Br.$$

(a) Regular Tetrahedron.

$$B = \frac{e^2}{4} \sqrt{3}, \quad r = \tfrac{1}{12} e \sqrt{6}, \text{ and } n = 4.$$

$$\therefore \text{ vol.} = \frac{4}{3} \cdot \frac{e^2}{4} \sqrt{3} \cdot \tfrac{1}{12} e \sqrt{6}$$

$$= \tfrac{1}{12} e^3 \sqrt{2}.$$

Cor. As the expression for the volume may be written $\frac{1}{3} \left(\frac{e}{\sqrt{2}} \right)^3$, therefore the cube on the side of a square whose diagonal is the edge of a regular tetrahedron is three times the tetrahedron.

(b) Regular Octahedron.

$$B = \tfrac{1}{4} e^2 \sqrt{3}, \quad r = \tfrac{1}{6} e \sqrt{6}, \quad n = 8.$$

$$\therefore \text{ vol.} = \tfrac{8}{3} \cdot \tfrac{1}{4} \cdot \tfrac{1}{6} \cdot e^3 \sqrt{18} = \tfrac{1}{3} e^3 \sqrt{2}.$$

Cor. This volume may be written $\tfrac{1}{6} (e \sqrt{2})^3$.

Therefore, the cube on the diagonal of a square whose side is the edge of a regular octahedron is six times the octahedron.

(c) Regular Dodecahedron.

By the methods of plane geometry, we find the area of a regular pentagon with side e to be

$$B = \tfrac{5}{4} e^2 \sqrt{\left(\frac{3 + \sqrt{5}}{5 - \sqrt{5}} \right)}.$$

Also, $\quad r = \sqrt{\left(\dfrac{25 + 11\sqrt{5}}{40}\right)}, \ n = 12.$

$$\therefore \text{vol.} = 5e^3\sqrt{\left\{\dfrac{3 + \sqrt{5}}{5 - \sqrt{5}} \cdot \dfrac{25 + 11\sqrt{5}}{40}\right\}}$$

$$= \dfrac{e^3}{4}(15 + 7\sqrt{5}).$$

(d) Regular Icosahedron.

$$B = \dfrac{e^2}{4}\sqrt{3}, \ r = e\sqrt{\left(\dfrac{7 + 3\sqrt{5}}{24}\right)}, \ n = 20.$$

$$\therefore \text{vol.} = \dfrac{20}{3} \cdot \dfrac{e^3}{4}\sqrt{\left(\dfrac{7 + 3\sqrt{5}}{8}\right)}$$

$$= \tfrac{5}{12}e^3(3 + \sqrt{5}).$$

EXERCISES H.

1. If a plane parallel to the bases, and midway between them, be passed through the prism of Art. 121, compare the areas of the sections of the three pyramids.

2. Apply the conditions of Ex. 1 to Art. 122.

3. A plane of section passes through the middle points of the parallel edges of a wedge, one of whose bases is a right section (Art. 124). Find the area of the section.

4. If e_1, e_2, e_3 be the three parallel edges of a wedge, show that $\frac{1}{3}(e_1 + e_2 + e_3)$ is the distance between the centroids of the bases.

5. Apply the prismoidal formula to find the volume of a wedge.

6. A prismoid has both bases parallelograms with angle θ, and the sides are a, b for the one, and a', b' for the other. Find its volume, its altitude being h.

7. Show that the cube on the side of a square whose diagonal is the edge of a regular octahedron is three-fourths of the octahedron.

8. If a regular tetrahedron and a regular octahedron have the same edge, the octahedron is four times the tetrahedron.

9. AA', BB', CC', DD', being diagonals of a cube, show that the plane through DBC' cuts off a pyramid whose volume is one-sixth that of the cube.

10. The direction edges of a cuboid are a, b, c, and a plane passes through the three distal extremities of these. Show that the area of the section is $\frac{1}{2}\sqrt{a^2b^2 + b^2c^2 + c^2a^2}$.

11. AA', BB', etc., are the diagonals of a ppd. Show that a plane through DBC' cuts off a pyramid which is one-sixth the ppd.

12. The direction edges of a ppd. are a, b, c, and the angles between them are $\angle(bc) = \lambda$, $\angle(ca) = \mu$, $\angle(ab) = \nu$. Then the vol. is

$$abc\sqrt{\{1 - \cos^2\lambda - \cos^2\mu - \cos^2\nu + 2\cos\lambda\cos\mu\cos\nu\}}.$$

OA, OB, OC are the direction edges;

$$\angle COB = \lambda, \; \angle COA = \mu, \; \angle AOB = \nu.$$

Let CP be normal to the plane of AOB, and PQ, PR be $\perp$s upon OA and OB.

vol. of ppd. $= OA \cdot OB \sin\nu \cdot CP$.

(P. Art. 215.)

$OQPR$ are concyclic, and OP is a diameter of the circumcircle;

$$\therefore \; OP = \frac{QR}{\sin\nu}, \qquad \text{(P. Art. 228.)}$$

and $\qquad CP = \sqrt{\left(OC^2 - \dfrac{QR^2}{\sin^2\nu}\right)} = \dfrac{1}{\sin\nu}\sqrt{(c^2\sin^2\nu - QR^2)}$;

$$\therefore \; \text{vol.} = ab\sqrt{(c^2\sin^2\nu - QR^2)}.$$

But $\qquad QR^2 = OQ^2 + OR^2 - 2OQ \cdot OR\cos\nu$;

and $\qquad OQ^2 = c^2\cos^2\mu, \text{ and } OR^2 = c^2\cos^2\lambda.$

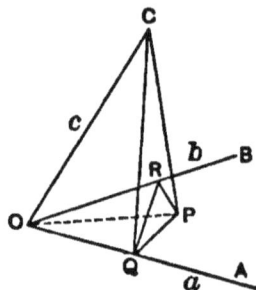

Whence, by substitution,

$$\text{vol.} = abc \sqrt{\{1 - \cos^2 \lambda - \cos^2 \mu - \cos^2 \nu + 2 \cos \lambda \cos \mu \cos \nu\}}.[1]$$

13. Show from the character of the result in Ex. 12 that if in any ppd. λ, μ, ν are all acute, all vertices except the one opposite O have one acute and two obtuse angles, etc.

14. With the vertices of a ppd. as centres, equal spheres are described to cut the ppd. Then the volume removed by all the spheres is equal to that of one of the spheres.

15. Show that space may be wholly divided up into regular octahedrons and tetrahedrons, and that there will be twice as many of the latter as of the former.

---

[1] The area of a parallelogram whose sides are a, b and angle θ is $ab \sin \theta$, or $ab\sqrt{(1 - \cos^2 \theta)}$, and the volume of the ppd. is $abc\sqrt{(1 - \cos^2 \lambda - \cos^2 \mu - \text{etc.})}$. On account of the analogy in form, the expression $1 - \cos^2 \lambda - \cos^2 \mu - \text{etc.}$ is sometimes called the square of the sine of the solid angle $O \cdot ABC$, and it usually appears in the matrix form

$$\begin{vmatrix} 1 & \cos \lambda & \cos \mu \\ \cos \lambda & 1 & \cos \nu \\ \cos \mu & \cos \nu & 1 \end{vmatrix}.$$

The analogy, however, is one of form only, as there are no functions of solid angles really corresponding to the sine, cosine, tangent, etc., of plane angles.

SECTION 2.

CONE, CYLINDER, SPHERE.

THE CONE.

131. The cone of Art. 67 is not a closed figure, and consequently does not admit of measurement for volume. But if the cone be cut by a plane which does not pass through the centre, and which makes, with the axis, an angle greater than the vertical angle, a closed figure is formed by the conical surface and the plane. It is this closed figure that is called a cone in relation to stereometry.

The centre of the cone is, in this relation, called the apex or vertex, and that portion of the section plane which forms a part of the enclosing figure is the base of the cone.

The word 'cone,' whenever having reference to stereometrical relations, will mean this figure.

132. As the director curve may be of any form, and as the plane of section may assume different relative directions, the variations in the cone are unlimited.

If the cone be circular, and the plane of section be perpendicular to the axis, the figure is the right circular cone; and this is the most important of all the cones.

The base is a circle, and the axis of the cone passes through the centre of the circle.

A right circular cone is generated by a right-angled triangle while revolving about one of the sides as an axis. The other side then generates the base (Art. 9. Cor. 1), and the hypothenuse generates the convex surface.

133. The cone may be looked upon as the limiting form of a right regular pyramid, when the number of sides in the base is indefinitely increased, and the length of each side is correspondingly diminished.

But the volume of any pyramid is one-third of its altitude multiplied by the area of its base;

Therefore, the volume of a cone is one-third of its altitude multiplied by the area of its base.

Cor. If the base be circular and its radius be r, its area is πr^2. And if h be the altitude of the cone, the

$$\text{vol.} = \tfrac{1}{3}\pi r^2 h.$$

134. The frustum of a cone is the limit of the frustum of a pyramid, and its volume is therefore

$$\tfrac{1}{3}h\,(B + B' + \sqrt{BB'}).$$

But if r and r' be the radii of the bases,

$$B = \pi r^2,\ B' = \pi r'^2,\ \text{and}\ \sqrt{BB'} = \pi r r'.$$
$$\therefore \text{vol.} = \tfrac{1}{3}\pi h\,(r^2 + r'^2 + rr').$$

THE CYLINDER.

135. When the cylinder of Art. 75 is cut by two parallel planes which cut completely through the surface, a closed figure is formed, which is the cylinder of stereometry.

When the planes are perpendicular to the axis of the cyliuder, the figure is a *right cylinder*. Otherwise it is an *oblique cylinder*.

136. It is obvious, from the definitions, that the cylinder is the limiting form of the prism, when the number of sides in the base is iudefinitely increased and the lengths of each side correspondingly diminished.

Hence the measure of a cylinder is the area of the base multiplied by the altitude (Art. 115. Cor. 2).

Cor. If the cylinder be circular aud right, and r be the diameter of the base,

$$\text{vol.} = \pi r^2 h,$$

where h is the altitude.

THE SPHERE.

137. *ABCD* is a tetrahedron in which the edge *AB* is equal and perpendicular to the edge *CD*, and *KJ*,

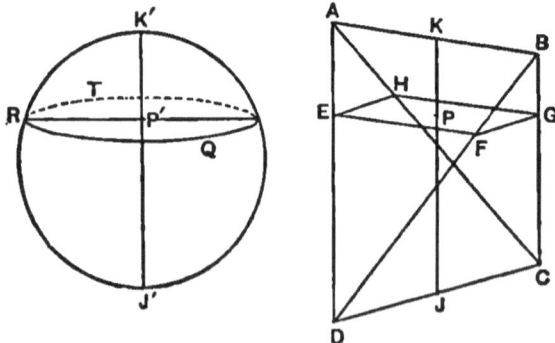

joining the middle points of these edges, is the common perpendicular to them.

Also, $K'RQT$ is a sphere having its diameter

$$K'J' = KJ.$$

We shall prove that corresponding laminæ of the tetrahedron and of the sphere are in a constant ratio, by proving that corresponding sections are in a constant ratio.

Proof. Let parallel planes pass through AB and CD. Then KJ is a common normal to these planes, and if the sphere be placed between the planes with $K'J'$ parallel to KJ, the planes will touch the sphere at K' and at J'.

Let the sphere and tetrahedron be relatively so placed, and let EG and RQT be corresponding sections of the figures (Art. 119).

Then $EFGH$ is a rectangle, and QRT is a circle, and $KP = K'P'$.

Now

$$\frac{KP}{KJ} = \frac{AE}{AD} = \frac{EH}{DC},$$

and

$$\frac{PJ}{KJ} = \frac{DF}{DB} = \frac{EF}{AB};$$

∴ by multiplication

$$\frac{KP \cdot PJ}{KJ^2} = \frac{EF \cdot EH}{AB^2} = \frac{\square EG}{AB^2}.$$

Also, denoting the radius of the sphere by r,

$$\frac{K'P' \cdot P'J'}{K'J'^2} = \frac{P'R^2}{K'J'^2} = \frac{1}{\pi} \cdot \frac{\pi \cdot P'R^2}{KJ^2} = \frac{1}{\pi} \cdot \frac{\odot QRT}{4r^2}.$$

Therefore ∵ $KP \cdot PJ = K'P' \cdot P'J'$,

$$\frac{\square EG}{\odot QRT} = \frac{AB^2}{4\pi r^2} = \text{a constant.}$$

Hence the corresponding sections of the tetrahedron and of the sphere are in a constant ratio; and the volumes of the tetrahedron and the sphere are in the same ratio.

Cor. 1. The tetr. : the sphere $= AB^2 : 4\pi r^2$.

But the tetr. $= \frac{2}{3} KJ \times$ mid. sec.   (Art. 130. *d.*)

$= \frac{1}{3} r \cdot AB^2$;

$\therefore$ vol. of sphere $= \frac{4}{3}\pi r^3$.

Cor. 2. The expression for the volume of a sphere may be written $\frac{2}{3} \cdot 2r \cdot \pi r^2$.

But $2r$ is a diameter of the sphere, and πr^2 is the area of a great circle. Therefore, $2r \cdot \pi r^2$ is the volume of the right circular cylinder which circumscribes the sphere.

Hence a sphere is two-thirds of its right circumscribing cylinder.

138. As the prismoidal formula applies to any portion of the tetrahedron confined between planes, each parallel to AB and CD, and since laminæ of the sphere hold a constant relation to corresponding laminæ of the tetrahedron, it follows that the prismoidal formula applies to any portion of the sphere limited between parallel planes.

Thus, applying the formula to the whole sphere, we have

$$B = 0, \ B' = 0, \ M = \pi r^2, \text{ and } h = 2r.$$

$$\therefore \text{vol.} = \frac{2r}{6}(0 + 0 + 4\pi r^2) = \frac{4}{3}\pi r^3.$$

139. *Def.* A portion of a sphere enclosed between two parallel planes is usually called a *zone* of the sphere; but if one of the planes is a tangent plane, the zone becomes a *segment* of the sphere.

140. *Volume of a zone.* Let a sphere be cut by parallel planes, given in section, in the diagram, by AB and CD; and let XY denote in section the plane which is parallel to the cutting planes, and half way between them.

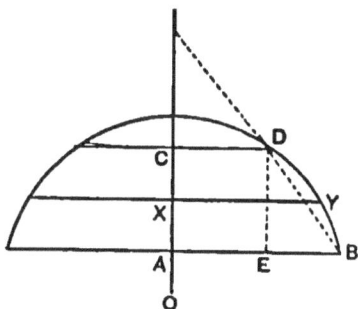

The data usually furnished from which to find the volume of the zone, are the radii CD and AB of the two bases, and the length of their common normal AC, or the altitude of the zone. Hence we suppose AB, AC, and CD to be the known quantities.

We have, O being the centre of the sphere,

$$OA^2 + AB^2 = OC^2 + CD^2 = OX^2 + XY^2,$$

since each expression is the square on the radius of the sphere.

$$\therefore OA^2 + AB^2 = (OA + 2AX)^2 + CD^2$$
$$= (OA + AX)^2 + XY^2.$$

Hence

$$AB^2 = 4AX^2 + CD^2 + 4OA \cdot AX$$
$$= AX^2 + XY^2 + 2OA \cdot AX.$$

$$\therefore 4OA \cdot AX = AB^2 - CD^2 - 4AX^2$$
$$= 2AB^2 - 2XY^2 - 2AX^2;$$

and hence

$$2XY^2 = AB^2 + CD^2 + 2AX^2.$$

Now, $\pi \cdot XY^2$ is the area of the middle section,

and $\pi \cdot AB^2$ and $\pi \cdot CD^2$ are the areas of the bases,

and $2AX$ is the altitude of the zone;

$$\therefore \text{vol.} = \tfrac{1}{6}\pi \cdot AC\{3AB^2 + 3CD^2 + AC^2\}.$$

Or, denoting the radii of the bases by r and r', and the altitude by h,

$$\text{vol.} = \tfrac{1}{6}\pi h\{3r^2 + 3r'^2 + h^2\}.$$

Cor. If $r = 0$, the zone becomes a segment, and its volume is $\tfrac{1}{6}\pi h(3r^2 + h^2)$.

141. The expression for the volume of the zone may be transformed as follows:

Draw $DE \perp$ to AB.

$$3AB^2 + 3CD^2 + AC^2 = 2(AB^2 + CD^2 + AB \cdot CD)$$
$$+ (AB - CD)^2 + AC^2;$$

and $\tfrac{1}{3}\pi \cdot AC(AB^2 + CD^2 + AB \cdot CD)$

is the volume of the frustum of the cone which has the same bases and altitude as the zone.

And AC being the projection of BD on OC, if we denote the angle between BD and AC by β,

$$AC = BD \cos \beta.$$

$$\therefore \tfrac{1}{6}\pi \cdot AC \cdot BD^2 = \tfrac{1}{6}\pi \cdot BD^3 \cos \beta$$

$$= \text{sphere on } BD \text{ as diameter} \times \cos \beta.$$

Therefore, the zone exceeds the inscribed conical frustum by the sphere on the slant height as diameter multiplied by the cosine of the semi-vertical angle of the cone.

EXERCISES I.

1. Compare the volume of a sphere (1) with that of the circumscribed cube ; (2) with that of the inscribed cube.

2. Compare the volume of the sphere with that of the circumscribed regular tetrahedron.

3. A cone circumscribes a sphere and has its slant height equal to the diameter of its base. Show that vol. of cone : vol. of sphere $= 9 : 4$.

4. If in Ex. 3 a plane passes through the circle of contact, the vol. of cone removed is $\frac{9}{8}$ the vol. of sphere removed.

5. A cylinder of radius a passes centrically through a sphere of radius r. Show that the volume removed from the sphere is $\frac{4}{3} \pi r^3 (1 - \cos^3 \theta)$, where $\sin \theta = \dfrac{a}{r}$.

6. A circular cone with semi-vertical angle a has its vertex at the centre of a sphere of radius r. Show that the volume common to the cone and sphere is $\frac{2}{3} \pi r^3 (1 - \cos a)$.

7. A right circular cone has its vertex lengthened out into a linear edge equal and parallel to a diameter of the base. Show that the volume is one-half that of the circumscribing cylinder. (The resulting figure is known as the common conoid.)

8. A cone whose semi-vertical angle is 45° has the diameter of a sphere as its axis, and its vertex on the sphere. Show that one-fourth of the sphere lies without the cone.

9. The cone of Ex. 8 has its semi-vertical angle equal to a ; then the part of the sphere lying without the cone is

$$\tfrac{1}{4} \pi r^3 (1 + \cos 2 a)^2.$$

142. In this section we propose, under three heads, *A*, *B*, and *C*, to explain and illustrate some special methods of measuring volumes, by applying these methods to the cone, cylinder, sphere, and some other spatial figures.

A. SPATIAL FIGURES GENERATED BY THE MOTION OF A PLANE FIGURE.

143. When a variable plane figure moves so that a fixed point lying in its plane describes a line or curve not complanar with it, the plane figure describes or generates a spatial figure.

The plane figure is then the *generator*, and the line or curve is the *path* of the particular point which describes it.

The case as here stated is too general for use, especially in elementary geometry or by elementary methods. We therefore subject the elements of the description to certain conditions, usually as follows.

(1) The generator is a closed plane curve, being invariable in form, while being either variable or constant in dimensions.

(2) The path is a line normal to the plane of the generator. This line will be called the *axis*.

(3) The generator preserves its orientation, *i.e.* any fixed line of the generator is invariable in direction; or

any fixed point in the generator describes a line or curve complanar with the axis. This line or curve, whose form depends upon the nature of the variation of the generator, is a guide to the motion of the generator, and forms the *director*.

Thus if the nature of the variation of the generator is given, the director is also given; and if the director is given, the nature of the variation is given.

144. Let PQR be a variable circle, whose centre, C, moves along the fixed line AB normal to the plane of the circle. AB is the axis.

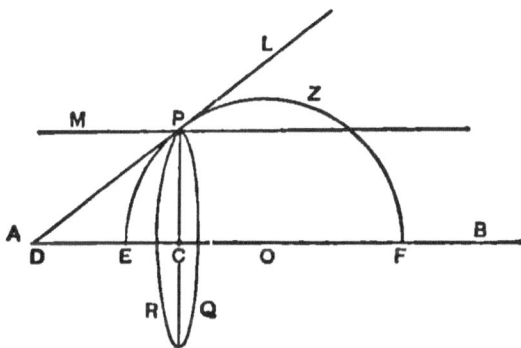

(1) Let P, any point on the circle, be guided by the fixed director line L, which meets AB in D.

Then, evidently the generating circle describes a cone having D as vertex and AB as axis.

The radius CP is in a constant ratio to DC.

Hence a variable circle, whose centre moves on a fixed line normal to its plane, and whose radius varies as the distance of the centre from a fixed point in the line, describes a cone.

(2) If the generating figure in case (1) were a polygon, the figure generated would be a right pyramid.

(3) Let P move on the line M parallel to AB.

The circle describes a cylinder, and a polygonal generator describes a prism.

(4) Let P be guided by the circle, Z, to which AB is a centre line, and EF a diameter.

The circle PQR then generates a sphere whose diameter is EF.

If O be the centre of the director circle, it is evident that $CP^2 + CO^2 = OP^2 = $ constant.

Therefore, a variable circle, whose centre moves on a line normal to its plane, and whose radius so varies that the sum of the squares on the radius and on the distance of the centre of the circle from a fixed point in the line is constant, generates a sphere.

(5) If the generator in case (4) were a polygon, the figure generated would be a polygonal *groin;* the most common groin is the square one.

In a similar manner many other figures may be generated, such as the oblate spheroid, the prolate spheroid, the hyperboloid, the paraboloid, the ellipsoid, etc.

145. Consider a number of equidistant points along the axis. Let the generator at these points be taken as bases of prisms or cylinders whose altitudes are the distances between consecutive points.

We have then a series of prisms or cylinders, of equal altitude, inscribed in or circumscribed about the spatial figure, as the case may be.

But (Art. 118) the volume of the spatial figure is the limit of either series of prisms or cylinders, when their number is indefinitely increased and their altitudes correspondingly diminished.

Hence if we can obtain an expression for the total volume of any number of such elementary prisms or cylinders, we can deduce the expression for the volume of the spatial figure, by imposing the condition that the number of elementary prisms or cylinders shall be infinite.

In carrying out this operation we assume the two following relations, which are proved in almost any work on algebra:

(A) $\qquad 1 + 2 + 3 + \cdots + n = \tfrac{1}{2}n^2 + \tfrac{1}{2}n$;

(B) $\qquad 1^2 + 2^2 + 3^2 + \cdots + n^2 = \tfrac{1}{3}n^3 + \tfrac{1}{2}n^2 + \tfrac{1}{6}n$,

where n denotes any positive integer, and the series extends from 1 to n.

146. Let X be a closed plane figure, which remains invariable in form while varying its dimensions.

Let a given point P be guided by the line AH, and let a point Q move on AC.

Then X describes a spatial figure, a cone or pyramid, having some position of the generator at B, as above.

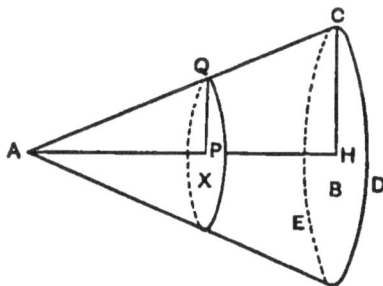

Let X denote the area of the variable figure, X, at any stage in its variation, and let B denote the area of CDE, the final stage of X.

Then $X:B = PQ^2:HC^2.$ (P. Art. 218. 5.)

And from similar triangles, APQ and AHC,

$$PQ^2:HC^2 = AP^2:AH^2.$$

$$\therefore \ X = \frac{AP^2}{AH^2} \cdot B.$$

Denote AH by h, and let AH be divided into n equal parts, and let AP be m of these parts.

Then $$AP = \frac{m}{n} \cdot AH,$$

and $$\therefore X = \frac{m^2}{n^2} \cdot B.$$

But the elemental cylinder, or prism, on X as base has $\frac{1}{n} \cdot AH$, or $\frac{h}{n}$, as its altitude, and therefore its volume is

$$Bh \cdot \frac{m^2}{n^3}.$$

This expresses the volume of any element, a particular one being got by giving a particular value to m. $m = 1$ gives the first element, lying next A; $m = 2$ gives the second, etc., and $m = n$ gives the last, lying next H.

The sum of these elements is

$$Bh \left\{ \frac{1^2 + 2^2 + 3^2 + \cdots + n^2}{n_3} \right\}$$

$$= Bh \left\{ \frac{1}{3} + \frac{1}{2n} + \frac{1}{6n^2} \right\}. \qquad \text{(Art. 145. B.)}$$

This holds true for all integral values of n. When we go to the limit by making n infinite, the fractions

$\frac{1}{2\,n}$ and $\frac{1}{6\,n^2}$ become zero, and the sum of the elements becomes the volume of the spatial figure (Art. 145).

$$\therefore \text{ volume} = \tfrac{1}{3}\,Bh.$$

As B may have any closed form whatever, this expresses the volume of any species of cone or pyramid which forms a closed spatial figure.

147. Let the generating figure, X, of constant form, but variable in dimensions, be guided by the axis OA, and by the circular quadrant CQA as a director, O being the centre of the quadrant.

Let CDE be the generator in the position in which O lies in its plane, and let S denote the area of CED, and X denote the area of the generator in any position.

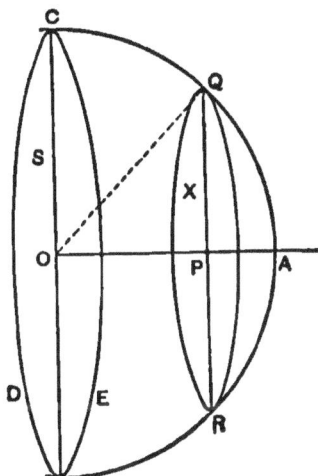

Then, since PQ is $\perp$ to OA,

$$PQ^2 = OQ^2 - OP^2 = OC^2 - OP^2.$$

But $\qquad X:S = PQ^2:OC^2;$ \qquad (P. Art. 218. 5.)

$$\therefore X = \frac{PQ^2}{OC^2}\cdot S = S - \frac{OP^2}{OC^2}\cdot S.$$

Now, denote OA by r and divide it into n equal parts, and let OP be m of these parts.

Then $\qquad OP = \frac{m}{n}r,$ and $OC = r.$

$$\therefore X = S - \frac{m^2}{n^2}\cdot S.$$

The elemental cylinder having X as base has $\dfrac{r}{n}$ for altitude, and its volume is therefore

$$rS\left(\frac{1}{n} - \frac{m^2}{n^3}\right).$$

The sum of these is

$$rS\left\{\frac{1+1+1+\cdots n \text{ terms}}{n} - \frac{1^2+2^2+\cdots n^2}{n^3}\right\}$$

$$= rS\left\{1 - \frac{1}{3} - \frac{1}{2n} - \frac{1}{6n^2}\right\}$$

$$= \tfrac{2}{3}rS$$

at the limit, when n becomes infinite.

And the volume generated while moving over the whole diameter is

$$\text{vol.} = \tfrac{4}{3}rS.$$

The value of this expression for volume depends upon the value of S.

1. If S is a circle, its area is πr^2, and the figure generated is the sphere.

$$\therefore \text{vol. of a sphere} = \tfrac{4}{3}\pi r^3.$$

2. If S is a square, and the middle point of its side is at C, the area is $4r^2$, and the figure is the common groin, and its

$$\text{vol.} = \tfrac{8}{3}r^3;$$

since the groin extends only from O to A.

3. If S is a regular hexagon with a vertex at C, we have a hexagonal groin, and its volume is $r^3\sqrt{3}$.

148. By varying the form of the generator, and also of the director curve, a great variety of spatial figures may be described.

1. With a circle as director and an ellipse as generator, QR being the major axis, we get the oblate spheroid; and with QR as minor axis, the prolate spheroid.

2. With an ellipse as director, and major axis as axis, and an ellipse as generator, we get the ellipsoid; with circle as generator we get the prolate spheroid.

3. With parabola as director, and a circle as generator, we get the paraboloid of revolution; and with ellipse as generator we get the elliptic paraboloid.

EXERCISES J.

The axes of an ellipse being a and b, its area is πab.

1. Show that the volume of a prolate spheroid is πab^2, where $a > b$.

2. Show that the volume of an oblate spheroid is $\pi a^2 b$, where $a > b$.

3. In the figure of Art. 147, if CQA were a quadrant of an ellipse, and $OA = a$ and $OC = b$, then $\dfrac{PQ^2}{OC^2} + \dfrac{OP^2}{OA^2} = 1$. Hence find the volume of an ellipsoid when the axes of. the generating ellipse are b and c at the position S.

4. In Art. 146, if $PQ^2 = c \cdot AP$, where c is a constant, show that the volume described is one-half that of the circumscribing cylinder.

5. OC is an axial line cut by a curve in O and C, and PM is a perpendicular from a point P on the curve to the axis OC. If $PM = a(OM \cdot OC - OM^2)$, show that the volume described by the curve in a revolution about the axis is $\frac{8}{15}$ of that of the circumscribing cylinder between O and C.

B. FIGURES OF REVOLUTION.

149. When a plane figure revolves about an axial line lying in its plane, the plane figure generates a spatial figure bounded wholly or partly by curved surfaces, and called, from its mode of generation, a *figure of revolution*.

Under the same circumstances the area of the plane figure generates a volume of revolution, *i.e.* the volume of the figure of revolution.

The area of a plane figure may be considered as the limit of the sum of a set of elements, composed of inscribed rectangles with equal but indefinitely small altitudes.

In revolution, these elements of area describe or generate elements of volume, whose sum has for its limit the volume of the generated spatial figure.

150. Let AC be a rectangle, and let it revolve about the axial line PR, parallel to AD.

The volume generated by the rectangle AC is the difference between the volumes generated by PC and by PD.

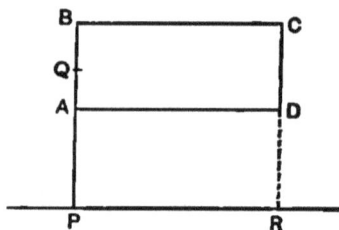

But   the vol. by $PC = \pi \cdot PB^2 \cdot BC$,

and   the vol. by $PD = \pi \cdot PA^2 \cdot BC$;

   $\therefore$ the vol. by $AC = \pi \cdot BC(PB^2 - PA^2)$

$$= \pi \cdot BC(PA + PB)(PB - PA).$$

If Q be the middle point of AB, PQ is the distance of the centre of the rectangle from the axis of revolution, and

$$PB + PA = 2\,PQ;$$

$$\therefore \text{ vol. by } AC = 2\pi \cdot PQ \cdot BC \cdot AB,$$

= area of $AC \times$ the circumference of the circle traced by the centre of AC.

Therefore, the volume described by a rectangle in one revolution about an axial line parallel to its side, and which does not cross the rectangle, is the area of the rectangle multiplied by the length of the path of its centre.

Cor. 1.   When the axial line passes through the centre of the rectangle, the length of path described by that centre is zero, and hence the volume described is zero.

From this it appears that if a revolving plane figure is crossed by the axis of revolution, the parts of the figure lying upon opposite sides of the axis generate volumes which must be taken in opposite senses, or with opposite signs.

Cor. 2.   From the figure we have $2\,PQ = 2\,PA + AB$; and hence $2\pi \cdot PQ = 2\pi \cdot PA + \pi \cdot AB.$

But when AC is an elemental rectangle, and we go to the limit by indefinitely diminishing AB, PQ has for its limit either PA or PB, these being finally the same. Hence, if the elemental rectangle AC is to be taken at the limit, PA may be taken for PQ.

151. Volume of a cone of revolution.   The rectangle AC revolves about AB as an axis.

The triangle ACB generates a cone of revolution; the rectangle generates a cylinder; and the triangle ACD describes that part of the cylinder which remains after the cone is removed.

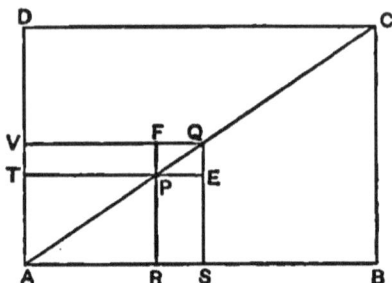

On AC take P, Q, any near points, which at the limit become coincident, and draw PR, QS, perpendicular to AB, and PT, QV, perpendicular to AD.

Then PS, being an elemental rectangle of the triangle ACB, generates an element of the cone; and PV, in like manner, generates an element of the portion of the cylinder which remains after removal of the cone.

But   vol. of element by $PS = \pi \cdot PR^2 \cdot PE$.

And   vol. of element by $PV = \pi(PR + FR)PT \cdot PF$.

And from similar triangles PRA and QEP,

$$\frac{PR}{RA} = \frac{QE}{EP}, \quad \text{or} \quad \frac{PR}{PT} = \frac{PF}{PE}.$$

$$\therefore \frac{\text{element by } PS}{\text{element by } PV} = \frac{PR}{PR + FR}.$$

And this relation being true for any, and therefore for every, pair of corresponding elements, is true for their sums.

But at the limit, when Q comes to P, PR and FR become the same.

$$\therefore \text{ the limit of } \frac{\Sigma(\text{elements by } PS)}{\Sigma(\text{elements by } PV)} = \frac{1}{2}.$$

Or,   cone by ABC : figure by $ACD = 1 : 2$.

Whence it follows that the cone is one-third of the cylinder.

REMARK. In the foregoing investigation we might, according to Cor. 2 of Art. 150, have taken the element described by PV as being $\pi \cdot 2\,PR \cdot PT \cdot PF$, since the element is finally to be taken at its limit.

152. Volume of a sphere. The quadrant DPA, and its circumscribed square $DBAC$, revolve about CA as an axis.

The quadrant generates a semisphere, and the square generates the right circumscribed cylinder.

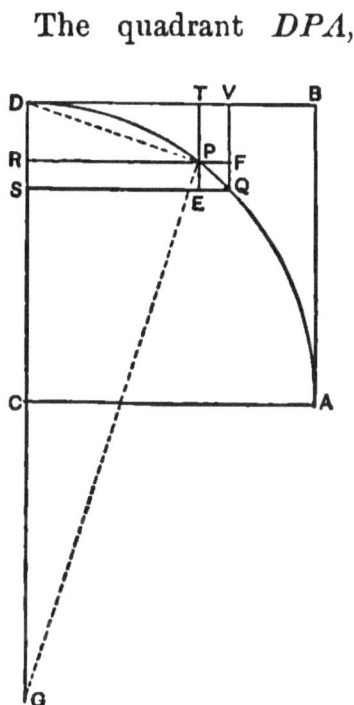

On the arc DA take P and Q, any near points which at the limit approach to coincidence.

Draw PR, QS, perpendiculars to CD, and PT, QV, perpendiculars to DB. Produce DC, making $CG = DC$.

The rectangle PS, being an element of the circle, describes an element of the sphere, and the rectangle PV for similar reasons describes an element of that part of the cylinder which lies without the sphere.

The volume of the element described by PS is, at its limit when Q comes to P, $2\pi \cdot CR \cdot PR \cdot RS$; and the volume of the element described by PV is, at its limit,

$$\pi(CR + CD) \cdot PT \cdot PF.$$

$\therefore$ at lt. $\dfrac{\text{element by } PS}{\text{element by } PV} = \dfrac{2 PR}{PT} \cdot \dfrac{RS}{PF} \cdot \dfrac{CR}{CR + CD}$.

But the $\triangle DPR$ and PGR being similar,

$$\frac{PR}{PT} = \frac{PR}{RD} = \frac{GR}{PR} = \frac{CR + CD}{PR}.$$

And the $\triangle PEQ$ and PRC being similar at the limit when Q approaches P,

$$\frac{RS}{PF} = \frac{PE}{PF} = \frac{PE}{EQ} = \frac{PR}{CR}.$$

$\therefore$ at lt., element by $PS = 2 \times$ element by PV.

And this being true for each, and therefore every pair of corresponding elements, is true for their sums.

Therefore the volume generated by the quadrant is twice the volume generated by the figure $DPAB$.

Or the volume of a sphere is two-thirds that of the circumscribing right cylinder.

Cor. 1. If r be the radius of the sphere, the volume of the circumscribing cylinder is $\pi r^2 \cdot 2r$; and hence the volume of the sphere is $\frac{4}{3} \pi r^3$.

Cor. 2. From the foregoing investigation it follows that wherever Q is taken on the arc, with CA as axis, the volume generated by the segment of the circle, $DSQP$, is two-thirds the volume generated by the rectangle $DSQV$.

153. Volume generated by an isosceles triangle revolving about an axis which passes through the vertex but does not cross the triangle.

The isosceles $\triangle OPQ$, with PQ as base, revolves about the axis OD passing through the vertex O.

Let PQ meet OD in D, and draw the altitude OR, and project P, R, Q, on OD at A, C, and B. Also draw QE parallel to OD.

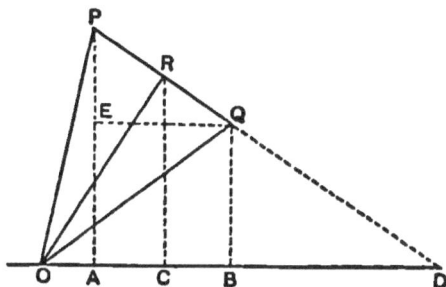

The $\angle APD = \angle ROD$,

and hence, $\qquad \triangle PEQ \backsimeq \triangle ORD.$

Therefore, $\qquad OD \cdot PE = OR \cdot PQ;$

also, $\qquad \triangle PEQ \backsimeq \triangle OCR,$

and $\qquad PQ \cdot CR = OR \cdot EQ;$

$\therefore OD \cdot CR \cdot PE = OR^2 \cdot EQ = OR^2 \cdot AB.$

Now, the vol. described by $\triangle OPQ =$ vol. of cone by $OPA +$ vol. of cone by $DPA -$ vol. of cone by OQB $-$ vol. of cone by DQB,

$$= \tfrac{1}{3}\pi \cdot OD\,(PA^2 - QB^2) = \tfrac{1}{3}\pi \cdot OD \cdot 2\,CR \cdot PE$$
$$= \tfrac{2}{3}\pi \cdot OR^2 \cdot AB.$$

Therefore, the volume described, in one revolution, by an isosceles triangle revolving about a line through its vertex, and lying without it, is the continued product of the projection of the base of the triangle upon the axis, the area of the square on the altitude, and the constant $\tfrac{2}{3}\pi$.

154. Let equidistant points A, B, C, etc., be taken in the arc of a circle of which O is the centre and OL is a centre line not crossing the arc.

The $\triangle AOB$, BOC, $\ldots$, are all isosceles and congruent.

The volume described by these triangles in revolving about OL as axis, p being the common apothem, is

$\frac{2}{3} p^2 \pi$ (pr. of AB on $OL +$ pr. of BC on $OL + \dots$).

But at the limit when the number of points A, B, $C \dots$ is indefinitely increased, and the distance between them is correspondingly diminished, the generating figure becomes the sector of a circle, p becomes the radius, and the sum of the projections of the bases of the triangles is the projection of the arc, and the figure generated is a sector of a sphere.

Therefore, the volume of a sector of a sphere $= \frac{2}{3} \pi r^2 \times$ pr. of the generating arc on the axis.

Cor. If the generating arc forms a semicircle, its projection on the axis is $2r$, and the figure generated is a sphere. $\therefore$ vol. of a sphere $= \frac{4}{3} \pi r^3$.

EXERCISES K.

1. Solve Ex. 6 of Set I., by the principle of 153.

2. AX is an axial line, and PM is a perpendicular to this line from a point P on a curve which starts from A. If $PM^2 = cAM$, where c is a constant, show that the volume described by one revolution about AX, is one-half that of the circumscribing cylinder.

3. The volumes of the circumscribing cylinder, the sphere, and the cone with the same base and altitude as the cylinder, are as the numbers 3, 2, and 1.

4. The volumes of the cylinder circumscribing a semisphere, the semisphere, and the cone with base and altitude of the cylinder, are as the numbers 3, 2, 1.

5. A plane cuts a sphere, and its circumscribed cylinder parallel to the base; then twice the volume of the segment is equal to the intercepted volume of the cylinder and twice the volume of the sphere on the altitude of the segment as diameter.

C. THEOREM OF PAPPUS OR GULDINUS FOR VOLUMES.

155. The mean centre of a system of complanar points for a system of multiples is defined (P. Art. 240) as the point of intersection of two lines, L and M, for which

$$\Sigma(a \cdot AL) = 0, \text{ and } \Sigma(a \cdot AM) = 0;$$

where A is a representative point, AL and AM representative perpendiculars from A to L and M respectively, and a a representative weight or number.

Also (P. Art. 241), if O be the mean centre of the system, and L be any line complanar with the system,

$$\Sigma(a \cdot AL) = \Sigma(a) \, OL.$$

We have to deal here with the mean centre of the area of a figure, and later on with the mean centre of the perimeter of a figure.

156. When a plane figure has an axis of symmetry, the mean centre of the figure lies on this axis.

For every point in the area upon one side of the axis of symmetry there is a point upon the other side exactly corresponding in every respect. So that if L be the axis of symmetry and A_1, A_2 be corresponding points, we have $A_1L + A_2L = 0$. And since the whole area is represented by pairs of such corresponding elements, $\Sigma(a \cdot AL) = 0$, or L passes through the mean centre.

Cor. 1. When a figure has two axes of symmetry, the mean centre of area is the point of intersection of the axes.

This is the case with the square, the rectangle, the rhombus, all regular polygons, the circle, and some other figures.

157. If the area of a figure be supposed to be made up of elemental squares, the centres of these squares, being their mean centre of area, will represent the points, A, B, C, etc., in a system of points, and the areas of the several squares will represent the weights.

But since the squares are all equal, the weights are all equal, and may be left out of consideration. With this understanding we have for the mean centre of area, $\Sigma(AL) = 0$, where L passes through this centre; and $\Sigma(AL) = n \cdot OL$, where O is the mean centre, L is any line not passing through O, and n is the number of elements under consideration.

158. Theorem. The join of the mean centres of two systems passes through the mean centre of the system composed of the two taken together.

This theorem is almost self-evident.

For if $\Sigma(a' \cdot A'L) = 0$ and $\Sigma(a'' \cdot A''L) = 0$ denote the two systems, and L is the join of their mean centres, we have at once

$$\Sigma(a' \cdot A'L + a'' \cdot A''L) = 0;$$

which is of the type $\Sigma(a \cdot AL) = 0$.

Cor. If any number of systems have their mean centres collinear, the mean centre of the system composed of all taken together lies on the line of collinearity.

159. Theorem. The mean centre of a parallelogram is its geometric centre, *i.e.* the intersection of its diagonals.

Let *ABCD* and *EFGH* be two congruent parallelograms, superposable with *E* on *A*, *F* on *B*, *G* on *C*, and *H* on *D*. Their mean centres of area are then coincident. But the parallelograms are also superposable with *E* on *C*, *F* on *D*, *G* on *A*, and *H* on *B*; and their mean centres of area are again coincident.

Hence the mean centre of each is the geometric centre.

Cor. In like manner it may be shown that when any figure has a geometric centre, that centre is also the mean centre of its area.

160. Mean centre of the area of a triangle. Let *BD* be a median to the triangle *ABC*. Draw *EF* and *GH* two near lines each parallel to *AC*, and draw *EI*, *FJ* parallel to *BD*.

The parallelogram *EJ* is an element of the area of the triangle, and the sum of the areas of these elements, when taken at the limit, is the area of the triangle.

But as *BD* bisects *EF* and *IJ*, it passes through the mean centres of all the elements of which *EJ* is a type.

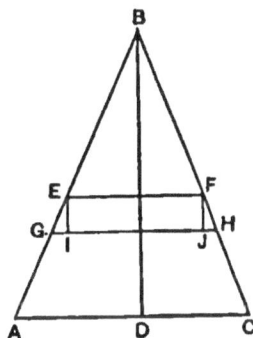

Therefore (Art. 158), the centre of area of the triangle lies on *BD*; and as it lies on both of the other medians, the centre of area of a triangle is its *centroid*.

Def. On account of the foregoing, we shall call the mean centre of area of any figure its *centroid*.

161. Theorem. The orthogonal projection of the mean centre of any complanar system is the mean centre of the projection of the system for the same multiples.

Let A, B, C, $\cdots$ be the elements in the plane U, and A', B', C', $\cdots$ be their projections on the plane V. Take L, any line in U, through the mean centre O, and let L' and O' be the projection of L and O on V.

Then $\qquad \Sigma(a \cdot AL) = 0.$ $\qquad$ (P. Art. 240.)

But AL, BL, CL, etc., are all parallel, and $A'L'$, $B'L'$, $C'L'$, etc., are all parallel. Therefore, the

$$\angle(AL \cdot A'L') = \angle(BL \cdot B'L') = \text{ etc.,}$$

and hence

$$\frac{AL}{A'L'} = \frac{BL}{B'L'} = \cdots = p, \text{ say};$$

$$\therefore \Sigma(a \cdot AL) = 0 = p\Sigma(a \cdot A'L'),$$

or $\qquad \Sigma(a \cdot A'L') = 0;$

and L' passes through the mean centre of the projected system. And as this is true for all directions of L and L' in their respective planes, O' is the mean centre of the projected system; or the projection of the mean centre of the system in U is the mean centre of the projected system in V.

162. *Def.* Let us call, in general, a figure of the type of the cylinder or prism, but with non-parallel bases, a *cylindroid*.

Suppose a system of near equidistant planes parallel to the axis to cut the cylindroid. These divide it into laminæ parallel to the axis. Now suppose a second set of planes, parallel to the axis, to cut the first system at right angles, and to have the distance between consecutive planes the same as in the first system.

These planes divide the cylindroid into elementary prisms on square bases. These form the *prismatic* elements of the figure, and the sum of their volumes, at the limit, as their bases are indefinitely diminished and their number is correspondingly increased, is the volume of the cylindroid.

163. AGB is a cylindroid having the base ABH normal to the axis, and the base CGD oblique to the axis.

Let PQ be a prismatic element, the area of whose base is β, and let the line CD be taken parallel to the common line of the planes of the bases, and let AB be the orthogonal projection of CD on the lower base.

Draw $QF \perp$ to AB, and FE normal to the base ABH. Then FE is parallel to QP, and meets CD in some point E. Draw $ER \perp$ to PQ. Join EP. Then EP is $\perp$ to CD, and $ERQF$ is a rectangle, and $EF = QR$.

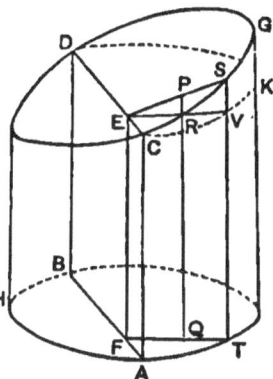

The volume of the prismatic element PQ is

$$\beta \cdot PQ = \beta \cdot PR + \beta \cdot RQ = \beta \cdot PR + \beta \cdot EF.$$

And since the bases of all the prismatic elements have the same area, and EF is constant, the sum of the volumes is

$$\Sigma(\beta \cdot PR) + \Sigma(\beta)EF.$$

But $\Sigma(\beta)EF$ is, at the limit, the volume of the cylinder HK, whose base is ABH, and altitude EF.

In order that this may be equal to the cylindroid, we must have $\Sigma(\beta \cdot PR) = 0$; and as every element PR is in a constant ratio to the corresponding element EP, and β is constant, we must have $\Sigma(EP) = 0$.

Or CD must pass through the centroid of the upper base, CGD, of the cylindroid; and (Art. 161) AB passes through the centroid of the lower base.

Hence, however the directions of the planes of section which give the bases may vary, provided they do not meet within the limits of the cylindroid, the volume remains unchanged, while the distance between the centroids of the bases remains the same.

Cor.   The volume of a cylindroid is the area of a right section multiplied by the distance between the centroids of the bases.

164. Let the plane figure X, invariable in form and dimensions, move from a position AB to another position CD, in such a manner that its direction of motion, whether following a line or a curve, is always normal to its plane.

Take two near positions of X as at GH and JK, and consider these as bases of a cylindroid forming an element of the figure generated by the motion of X. If P and Q be the centroids of the bases, the volume of the elementary cylindroid, GK, is $X \cdot PQ$, where X is the area of the generating figure.

And at the limit, when P approaches indefinitely to Q, the sum of the cylindroids is the generated spatial figure, and the sum of the elements PQ is the path of the centroid of X.

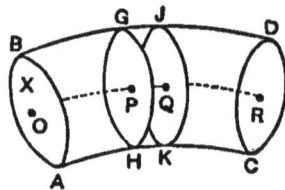

Therefore, when a plane figure, invariable in form and dimensions, moves in a path which is at every instant normal to the plane of the figure, the whole volume described is the area of the figure multiplied by the length of path moved over by the centroid of the figure.

This is the statement of the theorem as first given by Pappus (about 300), and afterwards reproduced by Guldinus (1577–1643), and usually called after his name.

Cor. When a plane figure revolves about a complanar axis, the direction of motion of the centroid is at all times necessarily normal to the plane of the figure, and the volume described in one revolution is the area of the plane figure multiplied by the circumference traced by its centroid.

Ex. A circle revolves about a complanar line lying without it; the figure generated is called an *anchor ring*. To find its volume.

Let r be the radius of the generating circle, and R be the distance of its centre from the axis. Then

$$\text{vol.} = 2\pi R \cdot \pi r^2 = 2\pi^2 r^2 R.$$

EXERCISES L.

1. Find the position of the centroid of a semicircle.

2. The circle which generates an anchor ring is divided by a diameter parallel to the axis; compare the volume described by the outer and the inner half of the circle.

3. A semicircle revolves about its limiting diameter. Any' segment whose chord is parallel to the axial line describes a volume equal to that of a sphere on the chord of the segment as diameter.

4. The distance from the centre of a circle to the centroid of any segment, is $\dfrac{2\,c^3}{3\,S}$, where c is the half chord of the segment, and S is its area.

5. The distance of the centroid of a segment from its chord is

$$\frac{2\,c^3}{3\,S} - \frac{c'^2}{2\,v} + v,$$

where c' is the chord of half the arc, and v is the versed sine of the arc (P. Art. 176. Cor. 1).

6. An arc of a circle revolves about its chord; the volume generated is

$$\frac{\pi}{3\,v}\{4\,c^3 v - 3(c^2 - v^2)S\}.$$

The figure generated is called a circular spindle.

7. The centroid of a semicircle is at the distance $\dfrac{4\,r}{3\,\pi}$ from the centre of the circle.

8. A semicircle revolves about a tangent at its middle point. The volume described is

$$\tfrac{1}{3}\,\pi r^3(3\,\pi - 4).$$

9. A square with side s revolves about a line through one vertex, making an angle θ with a side, and not crossing the square. The volume described is $\pi s^3(\sin\theta + \cos\theta)$.

10. A plane cuts through a right circular cylinder so as to cut one base only. The volume of the portion removed is

$$\frac{h}{v}\{\tfrac{2}{3}\,c^3 - (r - v)\,S\};$$

where h is the height of the convex part, r is the radius of the cylinder, and v, c, and S denote the versed sine, semichord, and area of the segment of the base.

This figure is called an *ungula* of a right circular cylinder.

SECTION 4.

PLANIMETRY — THE MEASUREMENT OF THE AREAS OF SURFACES, OR SUPERFICIES.

165. When a spatial figure is bounded by plane faces only, the area of its surface is the sum of the areas of its faces.

For such figures no special method is required outside of the processes of plane geometry.

The area of a curved surface is usually derived from that of a polyhedron by going to the limit, and supposing the number of polyhedral faces to be indefinitely increased while the size of each face is correspondingly diminished.

In some curved surfaces, however, we may suppose the surface to be brought to coincide with a plane by a sort of unrolling of the surface without stretching or distorting it in any of its parts. Such surfaces are said to be *developable ;* and when the surface is brought to coincide with a plane, it is said to be developed on the plane.

Thus a sheet of paper may be rolled into a cone or a cylinder, but it cannot be bent into a sphere.

The cylinder and the cone are accordingly developable surfaces, while the sphere is not.

It is readily seen that none but ruled surfaces can be developable. Ruled surfaces are not, however, all developable, and those which are not so are called *skew* surfaces.

166. Development of the conical surface.

Let O be the centre of a circular cone, and L be a generating line.

On L take any point, P, and through P draw the cone-circle APB with O as vertex.

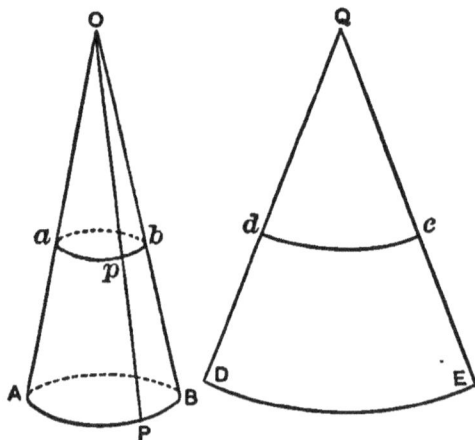

With any point, Q, as centre, and $QD = OP$ as radius, describe an arc, DE, equal in length to the circumference of the circle APB.

The figure QDE, a sector of a circle, is the development of the conical surface lying between the centre O and the cone-circle APB.

It must be remarked that the construction here given is theoretical only, since we have no method in elementary geometry of constructing an arc of one circle equal in length to a given arc of another circle, when the circles have different and incommensurable radii. This difficulty will not, however, vitiate any application to be made of this principle.

167. Area of surface of right circular cone.

For the closed cone $O \cdot APB$ it is evident that the area of the curved surface is equal to the area of its development, *i.e.* of the circular sector QDE, and this is one-half the length of the arc DE multiplied by the radius QD.

But the arc DE is equal in length to the circle APB, and the radius QD is equal to OP.

Therefore, denoting the circumference of the base by C, and the slant height, AO or BO, by S,

$$\text{curved surface} = \tfrac{1}{2} CS.$$

168. Frustum of a right circular cone.

Drawing a second cone-circle, apb, to the vertex O, and the development Qde, we have for the frustum,

$$\text{area} = \text{sector } QDE - \text{sector } Qde.$$

Or, denoting the circumference of apb by c,

$$2 \text{ area} = OP \cdot C - Op \cdot c.$$

But
$$OP = Pp + Op; \text{ and } \frac{OP}{Op} = \frac{C}{c};$$

$$\therefore \frac{Pp}{Op} = \frac{C - c}{c}, \qquad \text{(P. Art. 195. 1.)}$$

and
$$2 \text{ area} = Pp \cdot C + Op \, (C - c),$$

or
$$\text{area of surface} = \tfrac{1}{2} Pp \, (C + c).$$

169. In the cylinder, O, and therefore Q, goes to infinity, and QD and QE become parallel.

Hence DE and de become equal and parallel lines, and the development $DdeE$ is a rectangle.

Or, if h be the height of the cylinder, and r be the radius of the base,

$$\text{convex surface} = 2\,\pi r h.$$

170. Area of the surface of a sphere.

Let AB be a quadrant of a circle which generates a semisphere by revolving about OB as an axis.

Take two near points on the curve, P and Q, which at the limit come into coincidence, and draw the chord PQ. This chord describes the convex surface of a frustum of a cone, and the area of the surface is

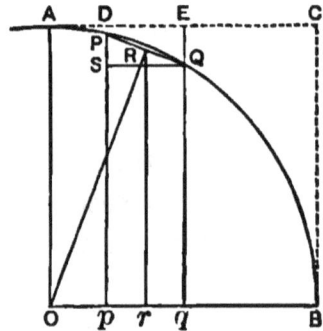

$$\tfrac{1}{2}PQ\,(2\,\pi\cdot Pp + 2\,\pi\cdot Qq), \qquad \text{(Art. 168.)}$$

where Pp and Qq are $\perp$s upon OB.

Take R, the middle point of PQ, and draw $Rr \perp$ to OB and join RO, and also draw $QS \perp$ to Pp.

Then the surface described by PQ is

$$2\,\pi\cdot PQ\cdot Rr.$$

And on account of the similar $\triangle$ PQS and ORr, the surface described by PQ is

$$2\,\pi\cdot OR\cdot pq.$$

And the convex surface described by a system of chords, forming the sides of a regular polygon, is

$$2\,\pi\cdot OR\cdot \Sigma\,(pq).$$

But at the limit when P comes to Q, the apothem, OR, becomes the radius, and the polygon becomes the circle; and the surface described by any arc, PQ, is

$$2\pi r \times \text{proj. of the arc on the axis.}$$

Now, $2\pi r$ is the circumference traced by D, a point on the circumscribed rectangle $ACBO$, and the projection of the arc is equal to DE;

Therefore, the convex surface described by the arc PQ is equal to the convex surface described by DE.

Hence, if a sphere and its circumscribed right cylinder be cut by two planes parallel to the bases of the cylinder, the area of the curved surface intercepted between the planes is the same for the sphere as for the cylinder.

Cor. The area of the surface of a sphere is equal to that of the curved surface of its circumscribed right cylinder.

Therefore, the area of the surface of a sphere is

$$4\pi r^2,$$

or four times the area of a great circle.

171. We may consider a sphere as circumscribing a conspheric polyhedron with an indefinite number of very small faces. Considering these faces as bases of pyramids having their vertices in common at the centre of the sphere, the sum of these pyramids at the limit, when the number is indefinitely increased, and the size of each base is correspondingly diminished, is the volume of the sphere, and the sum of their bases is the surface of the sphere.

But each pyramid is $\frac{1}{3} Bh$, and their sum is $\frac{1}{3} \Sigma (Bh)$. Or writing S for ΣB, and r for h,

$$V = \tfrac{1}{3} Sr;$$

where V is the volume of the sphere, and S is the area of the surface.

THEOREM OF PAPPUS OR GULDINUS FOR SURFACES.

172. For convenience we shall call the mean centre of the perimeter of a plane figure its *centre of figure*.

The general theorems respecting the mean centre as developed in Arts. 158 and 160 apply to the centre of figure in the same manner as to the centre of area.

173. Two equal line-segments, AB and CD, are congruent whether A is placed on C, and B on D, or A on D, and B on C, and hence the centre of figure of a line-segment is its middle point.

Then, in any rectilinear figure which has an axis of symmetry, the sides exist in congruent pairs which are symmetrically disposed upon opposite sides of the axis of symmetry.

And hence if A and A' denote the middle points of two sides forming a symmetrical pair, $AL = -A'L$, or $AL + A'L = 0$; where L is the axis of symmetry.

Therefore, $\Sigma (AL) = 0$; or L passes through the centre of figure.

Hence, when a rectilinear figure has an axis of symmetry, the centre of figure lies upon that axis.

Cor. When a rectilinear figure has two axes of symmetry, their point of intersection is the centre of figure.

174. Let *AGB* be a cylindroid having the base *ABH* normal to the axis, and the base *CGD* oblique to it. Suppose the convex surface to be divided in very narrow strips of equal width throughout, and parallel to the axis of the cylindroid and equal in width to one another; and let *b* denote the breadth of one of these elements of surface, and let *ST* denote the line along the middle of the strips. Then *ST* is normal to the base *ABH.*

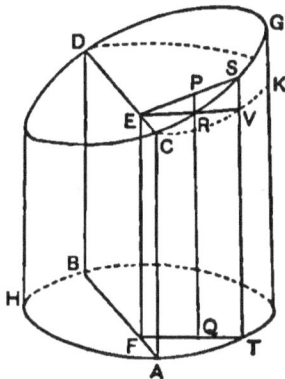

Let *CD* be parallel to the common line of the planes of the bases, and let *AB* be the projection of *CD* on the lower base. Draw $TF \perp$ to *AB* and $FE \perp$ to *CD.* Also draw $EV \perp$ to *ST*, and join *ES.*

Then, *EVTF* is a rectangle, and $VT = EF.$

The area of the element represented by *ST* is

$$b \cdot ST, \text{ or } b \cdot EF + b \cdot SV.$$

And the sum of these elements, of which the one represented by *ST* is the type, is,

$$\Sigma (b \cdot EF) + \Sigma (b \cdot SV).$$

But *EF* is constant, and $\Sigma(b)$ is the circumference of the base *ABH.*

Therefore $\Sigma(b \cdot EF)$ is the convex surface of the cylinder, or prism, whose base is *ABH*, and whose altitude is *EF*. And that this may be equal to the whole convex surface we must have $\Sigma(b \cdot SV) = 0$.

But as all the elements have the same width, b is constant, and $SV : SE$ is constant;

$$\therefore \ \Sigma(ES) = 0;$$

or, *CD* passes through the centre of figure of the upper base; and hence *AB* passes through the centre of figure of the lower base.

Therefore, the area of the convex surface of a cylindroid is the circumference of a right section multiplied by the distance between the centres of figure of the bases.

175. Let the plane figure X, invariable in form and dimensions, move with centre of figure on the path *OPQR*, and so that the direction of the path is at all points normal to the plane of the figure, and let *GH* and *JK* be two near positions, which at the limit come into coincidence.

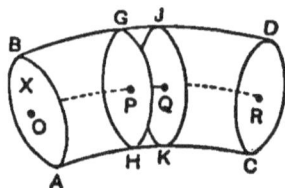

The surface of the elementary cylindroid *GK* is the circumference of $X \times PQ$; and the area of the surface of the figure generated by the motion of X is

$$\Sigma(PQ) \times \text{circum. of } X.$$

But $\Sigma(PQ)$ is the path *OPQR*⋯, and circumference of X is constant.

Therefore, the area of the surface described by a plane figure, invariable in form and magnitude, which moves so that its direction of motion is at each point normal to its plane, is the circumference of the generating figure multiplied by the length of path described by its centre of figure.

Cor. When a plane figure revolves about a complanar line as axis, the direction of motion is necessarily normal to the plane of the figure, and the surface described has for its area the circumference of the figure multiplied by the circumference traced by its centre of figure.

Ex. To find the surface of an anchor ring. The centre of figure is the centre of the generating circle, and the circumference traced is $2\pi R$.

$$\therefore \text{ area of surface} = 2\pi r \cdot 2\pi R = 4\pi^2 Rr.$$

176. The two theorems which go under the name of Guldin's theorems, but which were discovered by Pappus, express relations of the highest importance in mathematics both pure and applied. They enable us to find the centroid of a generating figure when the volume of the generated figure is known, or the centre of figure of a generating figure when the area of the surface of the generated figure is known, and *vice versa*.

Thus knowing the volume of a sphere, we can readily find the centroid of a semicircle, and knowing the surface of a sphere enables us to find the centre of figure of a semicircular arc.

EXERCISES M.

1. The circle describing an anchor ring is divided by a diameter parallel to the axis. Show that the difference between the surfaces described by the outer and the inner part is eight times the area of the generating circle.

2. The convex surface of a cone is πrs; and the entire surface is $\pi r(r+s)$; where s is the slant height, and r is the radius of the base.

3. The entire surface of a conical frustum is

$$\pi \{s\,(r+r_1) + r^2 + r_1{}^2\}.$$

4. The areas of the surfaces of the regular polyhedra are as follows:

Tetrahedron, $e^2\sqrt{3}$; Cube, $6\,e^2$; Octahedron, $e^2 2\sqrt{3}$; Dodecahedron, $e^2 15\sqrt{(1 + \tfrac{2}{3}\sqrt{5})}$; Icosahedron, $e^2 5\sqrt{3}$.

5. The distance between the centre of a circle and the centre of figure of any arc of the circle is $\dfrac{2\,cr}{l}$, where l is the length of the arc, and c and r denote as usual.

6. The area of the surface of a circular spindle is

$$\frac{\pi}{v}\left\{2\,cc'^2 - (c^2 - v^2)\,l\right\}.$$

7. The convex surface of an ungula of a right circular cylinder is $\dfrac{h}{v}(2\,cr + \overline{r - v}\cdot l)$. See Ex. 10. K.

Part IV.

———∘∘⟩∘⟨∘∘———

SECTION 1.

PERSPECTIVE PROJECTION.

177. *Def.* Let P be a variable point on a plane figure X; let O be a fixed point not complanar with X; and let L be the line OP.

When P describes the figure X, L describes the perspective projection of X in space, or the spatial projection of X, and O is the centre of the projection.

If the spatial projection be cut by a plane V, the figure of section is a plane figure called the *perspective projection* of X on V for the centre O.

When O goes to infinity, L has a fixed direction, and is parallel to a fixed line for all positions of P, and the projection becomes *parallel* projection.

If the direction of O at infinity is normal to V, the projection on V becomes *orthogonal* or *orthographic* projection, which is thus a special case of perspective projection.

· If the direction of O at infinity is perpendicular to

the plane of X, but oblique to V, the projection may be called the *ant-orthogonal* projection of X on V.

In what follows in this section, projection will mean projection with O finite unless otherwise stated, and *the projection* will mean the figure of section.

178. We observe that in a way projection and section are reciprocal processes, as by projecting a plane figure we get a spatial one, and by cutting the spatial one by a plane we return to a plane figure.

And this passing from one plane figure to another through a spatial figure may be repeated as often as we please.

179. Since the generator L is unlimited, the spatial figure extends to infinity on both sides of the centre, and admits of section on either side or section on both sides by the same plane, examples of which will occur hereafter.

180. The following theorems are fundamental:

1. A line projects into a line.

For the spatial projection of a line is a plane, and every plane section of a plane is a line.

2. The point of intersection of two lines projects into the point of intersection of the projections of the lines.

Hence the projection of a plane rectilinear figure is a plane rectilinear figure having the same number of sides and vertices as the projected figure.

3. A curve projects into a curve, and a tangent to the curve into a tangent to its projection.

For the coincident points common to the curve and
the tangent become coincident points common to the
curve and the tangent in the projection.

On account of its character, perspective projection is
also called conical projection.

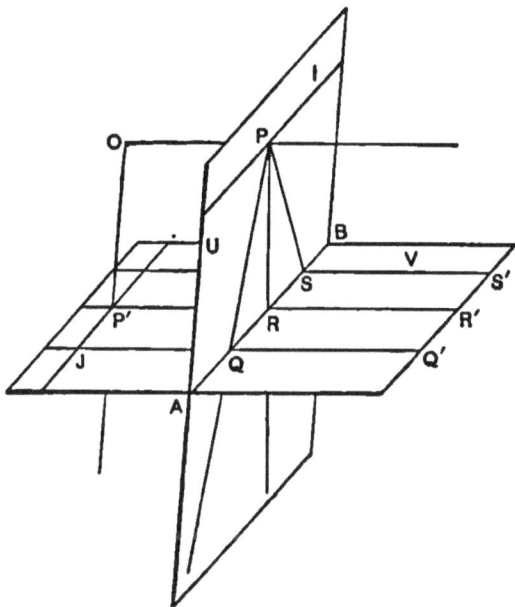

181. Let O be a centre of projection, and let U and V
be two planes, neither of which contains O, and whose
common line is AB.

Through O pass a plane parallel to V, and let it meet
U in I. Then I is parallel to AB. Take P, any point
in I, and let OP be the axis of an axial pencil. The
section of this pencil by U is the flat pencil $P \cdot QRS, \cdots$;
and since OP meets V at infinity, the section of the
axial pencil by V is a set of parallels, QQ', RR', SS',
etc., $\cdots$ (Art. 21. Cor. 2).

Hence the flat pencil on U with vertex at P projects into a parallel system on V, and the direction of the system is that of the axis OP.

Def. P is the *vanishing point* on U for the parallel system on V, or it is a vanishing point for V, *i.e.* for some system of parallels on V; and I, which is the locus of vanishing points for different systems of parallels on V, is the *vanishing line* on U, or the vanishing line for V.

Similarly, by passing a plane through O parallel to U, we obtain the line J, on V, as the vanishing line for U.

Thus either plane may be taken indifferently as the plane of the figure, and the other plane becomes the plane of the projection, or the plane of section. The operation of projection is thus completely reversible.

Cor. 1. Any point is projected to infinity by taking the plane of section parallel to the join of that point with the centre of projection.

Cor. 2. Any line is projected to infinity by taking the plane of section such that the given line may be the vanishing line for that plane.

Cor. 3. If P goes to infinity along I, OP becomes parallel to AB. But the flat pencil whose vertex is at infinity is a set of parallels.

Therefore, lines parallel to the common line of the planes project into lines having the same direction, *i.e.* into a set of parallels.

Cor. 4. If the planes U and V are parallel, the figure and its projection are similar (Art. 28. Cor. 2), and parallel lines project into parallel lines.

182. *Application.* To prove that the middle points of the three diagonals of a tetragram are collinear.

ABCD is a tetragram, and *AC*, *BD*, *EF* are its three diagonals.

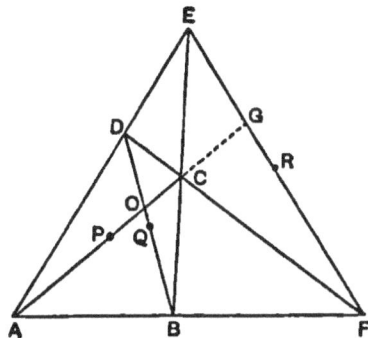

Let *BD* and *FE* produced meet in *S*, and project *S* to infinity in the direction *BD*. Then *BD* and *FE* become parallel, and *AC* is a median to the new triangle *AEF* (P. Ex. 11, p. 170), and *P*, *Q*, and *R* are points on this median, and are therefore collinear.

But by projection a line can come only from a line. Therefore, *P*, *Q*, *R*, are always collinear.

183. Theorem. Anharmonic relations are unchanged by projection.

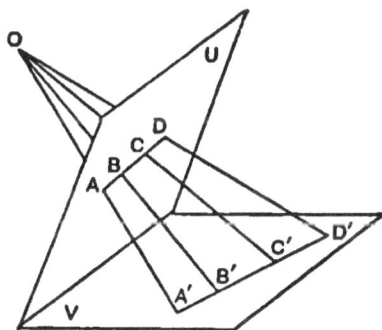

Let *ABCD* be a range on the plane *U*, and let *A'B'C'D'* be its projection on *V* from the centre *O*.

ABCD, *A'B'C'D'*, and *O* lie in one plane, and the theorem is reduced to that of a flat pencil.

But $O\{ABCD\} = O\{A'B'C'D'\}$. (P. Art. 304. Cor. 3.)

$$\therefore \{ABCD\} = \{A'B'C'D'\}.$$

Or the anharmonic ratio of *ABCD* is unchanged by projection.

Cor. 1.   Any given range can be considered as having come from some other range having the same anharmonic ratio; and any range and its projection are homographic.

Cor. 2.   If *ABCD* be a harmonic range, and *D* be projected to infinity, the projection of *B* bisects the join of the projections of *A* and *C*.

184. *Application.*   To show that a diagonal of a tetragram is divided harmonically by the other diagonals.

In the figure of Art. 182, project the diagonal *EF* to infinity (Art. 181. Cor. 2), *E* going in the direction *AD*, and *F* in the direction *AB*.

Then *ABCD* becomes a parallelogram, and *O* is the middle point of the diagonals, and therefore the middle point of *AC*.   But since *G* goes to infinity, *AOCG* is a harmonic range.

Therefore, *AOCG* is always a harmonic range.   Similar proofs may be obtained for the other diagonals.

185. Theorem.   Any angle less than a straight angle may be projected into any required angle less than a straight angle.

Let *APC* be the given angle lying in the plane *U*, and let *A* and *C* be the points where its arms meet the plane of projection *V*.

With *AC* as chord describe on *V* a segment of a circle *ABC* which shall contain the required angle, and through *B*, any point on this segment, draw *BP*.   The centre of projection is at any point on the line *BP*, as at *O*.

For the planes *BPA* and *BPC* contain the given angle *APC* on *U*, and the required angle *ABC* on *V*.

Cor. 1. Since *B* is any point on a circle, and *O* is any point on the line *BP, O* lies on the surface of a cone whose centre is *P*, and of which *OPB* is a generating line, and the circle *ABC* the director curve.

Cor. 2. Let *I* be any line in *U*.

Through *I* draw any plane, *W*, and take *V* parallel to *W*. From any point on *W*, *I* is projected to infinity on *V*. But *W* cuts the cone in a curve, *QOR*, and any point in this curve lies at the same time upon the surface of the cone and upon the plane *W*.

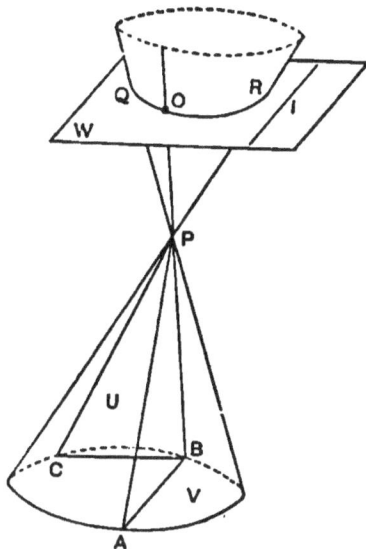

Therefore, from any point in this curve the ∠ *APC* is projected into a given angle *ABC*, and any line *I* is projected to infinity.

Hence any number of points can be found all lying on the plane section of a cone, from which as centre a given angle of a plane figure may be projected into a required angle, and a given line of the same figure be projected to infinity.

186. Any quadrilateral may be projected into a rectangle.

For if we project the external diagonal to infinity, and an angle of the quadrilateral into a right angle, the figure becomes a parallelogram with one right angle, *i.e.* a rectangle.

187. Theorem. Any line-segment may be so projected that any point on the line of the segment may become the middle point of the projection.

1. Let the point divide the segment internally.

Let AB be a given segment and C be an internal point in it. Take D, the harmonic conjugate to C (P. Art. 309), and taking any point O as centre of projection, join OD, and project the range upon a line, L, parallel to OD.

Then since D goes to infinity, C' bisects the segment $A'B'$.

2. Let the point D divide the segment externally. Take C, a harmonic conjugate to the given point D, and joining OC, project the range upon any line M parallel to OC.

Then, since C goes to infinity, D'' bisects the segment $A''B''$.

In this projection we notice that no part of the projected segment lies between A'' and B'' in the finite, but that the projection of the line-segment AB extends from A'' upwards to C'' at infinity, and thence returns, from below, to B''.

We have thus reversed the segments of the original line, so that the finite part ACB extends through infin-

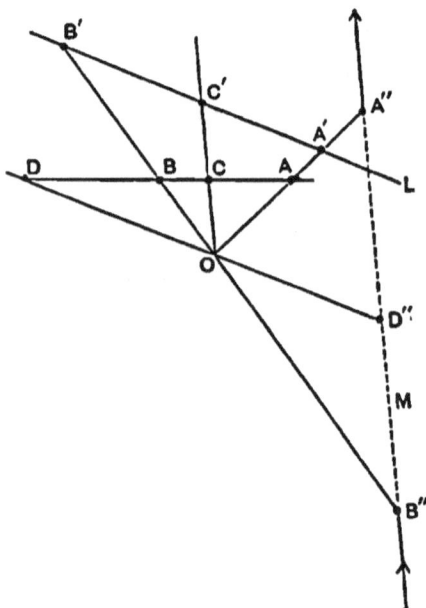

ity, and the infinite part *BDA* becomes finite; and thus *D″* represents the external point of bisection of *AB*, that is, the point at infinity, projected into the finite. We have thus a graphic illustration of the theorem that the point at infinity on any line-segment is the external point of bisection of the segment.

Cor. 1. A circle may be so projected that any point within it may become the centre of the curve of projection.

Let *C* be the given point in the circle, and let *ACB* be a diameter, and *ECF* a chord perpendicular to this diameter.

On any plane parallel to *ECF* project the segment *ACB* so that *C′* may be the centre of *A′B′*, by 1.

In this case no part of the circle goes to infinity, and the projection is a closed curve.

Cor. 2. A circle may be so projected that any point without it may become the centre of the figure of projection.

Apply the principle involved in 2.

In this case the circle becomes two curves which extend to infinity in opposite directions, and which do not meet one another. The diameter *AB* projects into a common axis to the two curves, and the projection of the given point bisects the part of the common axis intercepted between the curves.

EXERCISES N.

1. Show that a tetragram may be projected into a rectangle.

2. A given line may be projected to infinity, and two given angles be projected into required angles.

3. ABC is a triangle, and DE is parallel to AC, D being on BC, and E on AB. CE and AD intersect in O. Then BO is a median; and if BO meets DE in P and AC in Q, $BPOQ$ is a harmonic range.

4. ABC is a triangle, and AD, BE and CF are parallel, D being on BC, F on AB, and E on AC. Show that

$$AE : EC = AF \cdot BD : BF \cdot CD.$$

5. A chord of a circle is projected to infinity. Show that the pole of the chord becomes the point of intersection of tangents which touch the projection at infinity.

SECTION 2.

PLANE SECTIONS.

188. The definition of a plane section, and some general facts with regard to it, are given in an earlier portion of this work (see Arts. 19 *et seq.*).

Evidently the plane section of any polyhedron is a polygon, and the plane section of a sphere is a circle. Such sections offer no distinctive features other than what belong in general to polygons and circles.

But when we make plane sections of the cone or cylinder, we are introduced to curved figures which are not circles, and with which we have not hitherto become acquainted.

These we propose to consider.

PLANE SECTIONS OF THE CIRCULAR CONE.

189. The spatial projection of a circle, from a centre of projection for which the circle is a cone-circle, is a circular cone. The section of this cone by a plane, variable in direction, is a variable curve, which, in passing through several distinctive phases in its variation, constitutes a class of plane curves which are known as *conic sections*, or simply *conics*.

Hence a circle may be projected into any conic; and conversely, any conic can be projected into a circle. And thus any conic may be projected into any other conic.

CLASSIFICATION OF CONICS.

190. Let O be the centre of the circular cone $O \cdot AKB$, L being a generating line, and let V denote a plane of section passing through any point Q. Then,

1. When V is normal to the axis of the cone, the section is a circle, DCQ, or C.

Now, let a tangent to the circle, C, at Q be drawn in the plane V, and let the plane revolve about this tangent line as an axis. Then,

2. When V makes with the axis of the cone an angle less than a right angle, and greater than the semivertical angle of the cone, the section is an *Ellipse, E.*

In this case the section-plane, V, cuts completely through one nappe of the cone, and does not meet the other nappe. Hence the ellipse consists of a single closed curve, as represented by the figure E.

3. When V makes with the axis of the cone an angle equal to the semivertical angle of the cone, the section is a *Parabola, P.*

In this case V is parallel to a single generating line, and cuts only one nappe of the cone, but does not cut through it, and thus the curve extends indefinitely outwards in one direction.

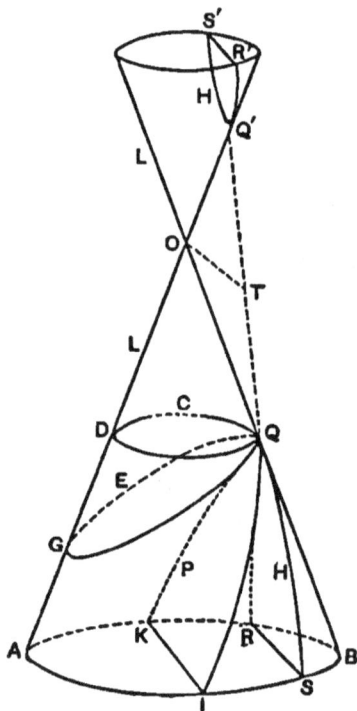

The parabola is consequently a single curve, but not a closed curve.

4. When V makes with the axis an angle less than the semivertical angle of the cone, the section is a *Hyperbola*.

In this case V is parallel to two generating lines which make a finite angle with one another, and cuts into both nappes of the cone, but does not cut through either.

Therefore, the hyperbola, H, consists of two curves, or rather two branches extending infinitely outwards in opposite directions, and separated from one another by a finite interspace, QQ'.

191. *Degraded forms.* All the conics may take what are called degraded forms; that is, forms which they assume as limiting forms, under the sequence of variation, but which are not visible curves.

1. Suppose that while the different directions of the section plane, which give the several conics, remain the same, the section plane moves up to O.

Then, (*a*) The circle and ellipse reduce to a point called a *point-circle* and a *point-ellipse* respectively. Of course there is no final distinction between a point-circle and a point-ellipse, but their names indicate their origin.

(*b*) The plane of the parabola becomes a tangent plane to the cone, and touches the cone along two coincident lines; and thus the parabola degrades into two coincident lines.

(*c*) The plane of the hyperbola gives in section two generating lines which make a finite angle with one another, and the hyperbola thus degrades into a pair of intersecting lines.

Hence a pair of intersecting lines is frequently called a *rectilinear hyperbola*.

Let V be the plane which gives the hyperbola II, H (Fig. 190), and let V' be parallel to V and pass through O. V' gives in section the rectilinear hyperbola which corresponds to H, H; and if we draw OT to the middle point of QQ', and by parallel projection in the direction TO, we project the hyperbola II, H upon the plane V', we have, on that plane, a hyperbola and its corresponding rectilinear hyperbola. The two lines which form the latter are then called the *asymptotes* of the former.

2. If G remains fixed while Q moves up to O, the ellipse becomes a double line-segment, and is called a *line-ellipse*.

192. From the generation of the various conics, as now explained, we deduce —

1. That the circle is a special form of the ellipse, and that properties of the circle are special cases of more general properties belonging to the ellipse. And as only one direction of the plane of section, relatively to the axis of the cone, can give the circle, the circle has only one form, or all circles are similar to one another.

2. That the parabola stands intermediate between the ellipse and the hyperbola, and is the form through which one of these curves passes into the other. Also, since only one direction of the plane of section relatively to the axis of the cone can give the parabola, the curve has only one form, and all parabolas are similar to one another.

3. That both the ellipse and the hyperbola are variable in form, the ellipse varying from the circle at the one limit to the parabola at the other, and the hyperbola varying from the parabola at the one limit to the form of two coincident lines at the other.

4. That all the conics have many properties in common.

COMMON PROPERTIES OF CONICS.

193. As a line can meet a circular cone but twice (72), so a line can meet a conic section in two and only two points. When these two points become coincident, the line becomes a tangent line, and the point of contact is a double point.

The conics constitute a distinct class of curves. Being the simplest curves that it is possible to have, they are called curves of the first order.

All curves not conics belong to a higher order, and cannot be obtained as sections of a circular cone by a plane, nor as sections of any cone, one of whose sections is a conic.

Curves are classified according to the number of times they may be met by a line under the most favourable circumstances, and all curves other than conics can be met by a line in more than two points, either real or imaginary.

Cor. A tangent to a conic lies completely without the conic, except at the point of contact.

194. *Z* is a sphere, and *O · PFG* is a tangent cone touching the sphere in the small circle *BEC*.

Denote the plane of *BEC* by *U,* and let *V* be a plane of section of the cone, touching the sphere in *S,* and meeting *U* in the line *DH.*

Let *W* denote the plane containing the axis of the cone and being normal to *DH.* Then *W* is perpendicular to both *U* and *V.*

AQP is the conic formed by *V,* and *P* is any point on this conic. *BD* is a centre line of the circle *BEC,* and is the common line of *U* and *W,* and *SA* is the common line of *W* and *V.*

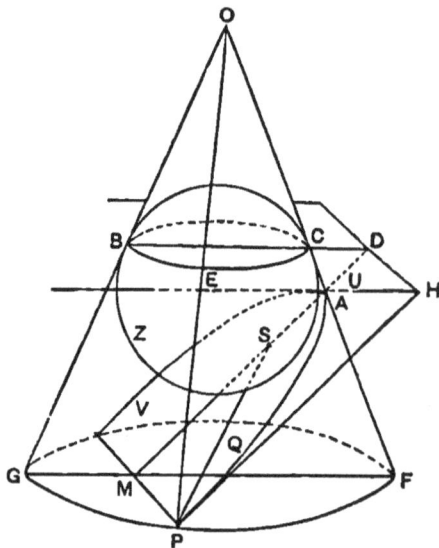

This latter line, *SA,* is the *principal axis* of the conic, or simply the axis, and it is perpendicular to *DH.*

The conic is evidently symmetrical about the axis *SA,* or the principal axis.

Draw *PH* ⊥ to *DH* and *PM* parallel to *DH.*

Then, *P, M, D, H* are complanar, all lying in *V,* and *MH* is a rectangle, and therefore *PH = MD.*

Join *PS* and *PO,* and pass a plane through *P,* parallel to *U,* giving the circular section *GPF,* and meeting *W* along the line *GF.*

Then *BEC* and *GPF* being cone-circles with *O* as vertex,

$$OP = OF, \text{ and } OE = OC;$$

$$\therefore CF = EP = SP. \qquad \text{(Art. 86. 1.)}$$

Similarly, $SA = CA.$

But, from similar triangles CAD and FAM,

$$CF : MD = CA : AD,$$

or $$SP : PH = SA : AD.$$

But the ratio $SA : AD$ is independent of the position of P on the conic, and remains constant while P moves along the conic.

Therefore, denoting this constant by e, we have

$$\frac{SP}{PH} = e = \text{a constant.}$$

Hence a conic, considered as the locus of a variable point, may be defined as follows:

A conic is the locus of a point which, being confined to one plane, so moves that its distance from a fixed point (S) is in a constant ratio (e) to its distance from a fixed line (DH), all being complanar.

This definition is usually adopted in analytical conics, and it is sufficiently general to include every conic.

Def. The point S is the *focus*, and the line DH is the *directrix*. A is the *vertex* of the conic, and the constant e is the *eccentricity*.

PM is an *ordinate* to the principal axis, and PS is the *focal distance* of the point P.

195. Let the accompanying figure represent the section by the plane W.

A second sphere, Z', may be drawn, to touch the cone and the plane V at S'. Then BC and $B'C'$ representing the sections of the circles of contact, the planes of these

circles are parallel, and they cut the plane V in parallel
lines represented in section at D and D'. Hence the
conic has two foci, S and S', two vertices, A and A', and
two parallel directrices represented in section by D
and D'.

The figure as here drawn applies particularly to the
ellipse, but it may serve as a type for all the other conics.

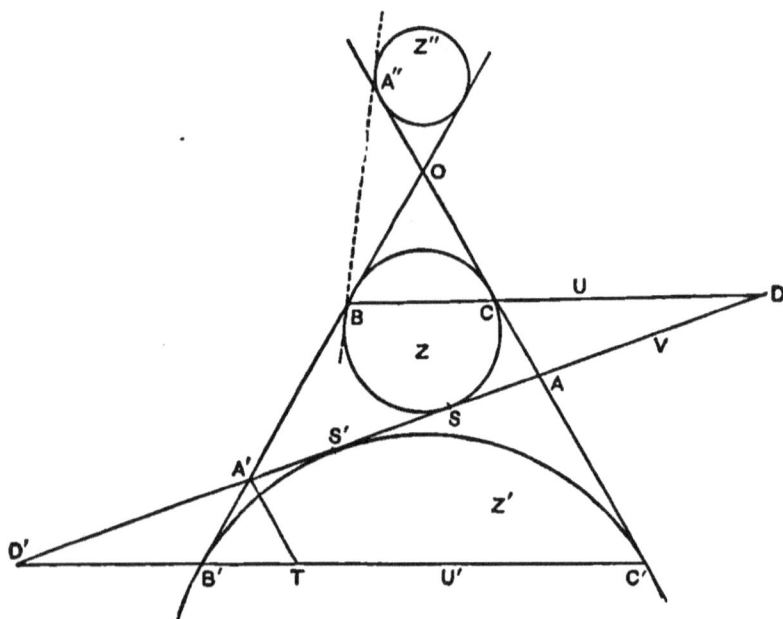

In the parabola one vertex, focus, and directrix are at
infinity.

In the hyperbola the second focus is given by the
point of contact of the sphere Z''.

Thus in the ellipse the curve lies between the directrices, while in the hyperbola the directrices lie between
the vertices of the two branches of the curve.

196. In the figure of Art. 195, considered as a plane figure, Z is the incircle, and Z' is an excircle of the triangle AOA'. Therefore, $AS = A'S'$ (P. Art. 135. Ex. 1); so that both foci are similarly situated with respect to the corresponding vertices.

Also, drawing $A'T$ parallel to CA,

$$A'S' = A'B' = A'T = AS = AC.$$

Therefore, the triangles $D'A'T$ and DAC are congruent, and $A'D' = AD$; or the directrices are similarly situated with respect to the corresponding foci and vertices.

Hence it follows that the same curve may be drawn from either vertex with the corresponding focus and directrix, and therefore that a line drawn at right angles to the principal axis, and bisecting the distance between the foci, is an axis of symmetry of the curve.

And as the principal axis is also an axis of symmetry, a conic has, in general, two axes of symmetry bisecting one another at right angles. These are called *the axes* of the conic.

In the parabola one axis is at infinity.

197. The character of the conic is determined by the value of its eccentricity.

In the figure to Art. 194,

$$e = SA : AD = CA : AD.$$

1. Let AD be infinite. Then $e = 0$, and the plane V is parallel to the plane U, and the conic is a circle.

Therefore, the eccentricity of a circle is zero.

· In this case both spheres touch V at the centre of the

circular section, and the foci of the circle become coincident at the centre of the circle.

2. Let $AD = CA$. Then $e = 1$; and the triangle CAD being isosceles,

$$\angle ADC = \angle ACD = \angle BCO = \angle OBC.$$

Therefore, AD is parallel to OB, and the conic is a parabola.

Hence the eccentricity of the parabola is unity. In this case a second sphere cannot be drawn in any finite position so as to satisfy the conditions for a focus.

3. When AD is $< \infty$ and $> CA$, the value of e lies between zero and unity. The $\angle ACD$ is $>$ the $\angle ADC$, and the plane V, being inclined to the axis of the cone at an angle greater than the semivertical angle of the cone, cuts through one nappe and gives the ellipse.

4. When AD is $< AC$ and > 0, e lies between unity and infinity; the angle ADC is greater than ACD, and the plane V, being inclined to the axis of the cone at an angle less than the semivertical angle of the cone, cuts into both nappes and gives the hyperbola.

198. When the centre of a circular cone goes to infinity, the cone becomes a cylinder, and the only possible plane sections are the circle, the ellipse, and two parallel lines representing the parabola and hyperbola.

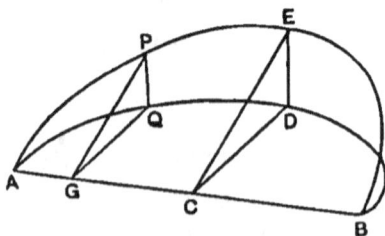

The ellipse, including the circle as a particular case, is the most important of the conic sections, and we propose to develop some of its more prominent properties, through its relationship to the circular cylinder.

Let ADB be one-half of the right section of a circular cylinder, and let AEB be the corresponding half of an oblique section of the same cylinder. Then ADB is a semicircle with AB as diameter, and AEB is one-half of an ellipse.

From C, the centre of both the circle and the ellipse, draw CD and CE perpendicular to AB, the former in the plane of the circle, and the latter in that of the ellipse. Then ED is a segment of a generating line of the cylinder.

On the ellipse take any point P, and draw PQ parallel to ED, and QG parallel to DC.

The $\triangle ECD$ and PGQ are similar, and $CD = CQ =$ the radius of the cylinder, and CD is $> GQ$.

Therefore, ED is $> PQ$.

But $CE^2 = CD^2 + DE^2 = CQ^2 + DE^2$, and this is $> CQ^2 + PQ^2$.

Whence CE is $> CP$.

Or, CE is the longest segment from C to the ellipse, and is the *semiaxis-major* of the ellipse.

In like manner it may be shown that CP is $> CA$; or that CA is the shortest segment from the centre to the ellipse, and is the *semiaxis-minor* of the ellipse.

These axes are perpendicular to one another.

199. On account of the similar triangles, PGQ and EDC (Fig. of 198),

$$PG : GQ = EC : CD.$$

But $EC : CD$ is constant for a constant direction of the plane of oblique section.

$$\therefore PG : GQ = \text{a constant.}$$

Hence the following construction for an ellipse; AQD is a quadrant of a circle with centre C, and GQ is a chord $\perp$ to AC. Take GP, a constant multiple of GQ. The locus of P is an ellipse whose semiaxis-minor is AC.

Again, draw $PH \perp$ to CE and let CQ meet HP in R.

Then, from similar triangles RPQ and CGQ,

$$\frac{RP}{GC} = \frac{PQ}{GQ};$$

$$\therefore \frac{RH}{PH} = \frac{PG}{QG} = \text{constant.}$$

And PH is a constant part of RH; hence the theorem —

If on a chord of a circle, perpendicular to a fixed diameter, a point be taken so as to divide the chord in a constant ratio, the locus of the point is an ellipse, and the fixed diameter is the major or the minor axis of the ellipse, according as the point divides the chord internally or externally.

Def. The circles which have the major and minor axes of the ellipse as their diameters are the major and minor *auxiliary circles* to the ellipse.

200. In the figure of Art. 198, let AEB be a semicircle with AB as diameter and C as centre, and let CE be perpendicular to AB.

Project orthogonally on any plane, V, passing through AB. Then $GQ = GP \cos PGQ$, and as PGQ is a con-

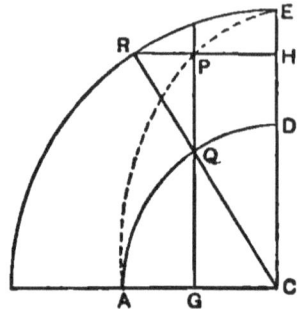

stant angle, GQ is in a constant ratio to GP and is less than GP.

Hence Q lies on an ellipse of which AB is the major axis.

Therefore, the orthogonal projection of the circle on any plane not parallel to its own is an ellipse.

This result also follows directly from the statements of Art. 190, since, in general, the perspective projection of a circle on any plane which cuts through one nappe is an ellipse. But in the case of perspective projection the same diameter of the circle does not project into an axis of the ellipse.

201. Conjugate diameters. Let $ADBE$ be a right section of a circular cylinder, by the plane U, and let $adbe$ be an oblique section by the plane V. Also let AB and DE be perpendicular diameters of the circle, and FG be a chord parallel to AB.

Now let the whole figure on U be projected ant-orthogonally (Art. 177) on V.

Then c is the centre of the ellipse and ab and de, the projections of AB and DE, are a pair of *conjugate diameters* of the ellipse,

And FG is bisected at H. Whence, from the nature of parallel projection, fg is bisected at h, and is parallel to ab.

Therefore, de bisects all chords parallel to ab; and in like manner ab bisects all chords parallel to de.

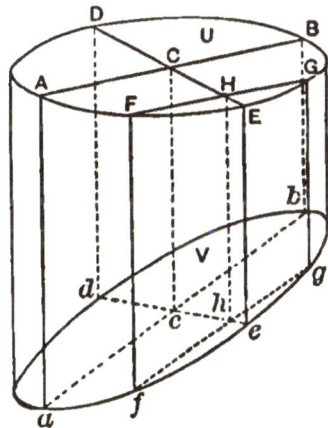

Hence, when two diameters are conjugate, each bisects all chords parallel to the other.

Cor. 1.   Conjugate diameters are such that each is parallel to the tangents at the extremities of the other.

Cor. 2.   Conjugate diameters in the ellipse are the parallel projections of orthogonal diameters in that circle whose projection gives the ellipse.

Manifestly an indefinite number of pairs of conjugate diameters may be found, and each diameter has one, and only one, conjugate.

Cor. 3.   When the plane V is parallel to AB or DE, *ab* and *de* are perpendicular to one another, and become the principal diameters of the ellipse.

202. Since Aa, Bb, etc., are generating lines of the cylinder, they are lines of contact of tangent planes (Art. 73).   Let four tangent planes to the cylinder touch it at Aa, Bb, Dd, and Ee.   The section of these by U is a square, whose sides are parallel to AB and DE; and this section remains constant in area however AB and DE may be drawn, provided they are diameters which intersect orthogonally.

The section of the tangent planes by V is a parallelogram whose sides are parallel to *ab* and *de*.

Now (Art. 116), the area of the square is equal to that of the parallelogram multiplied by the cosine of the angle between U and V.

Therefore, the area of the parallelogram is unaffected by any change in the position of V, which does not change the inclination of that plane to the axis of the cylinder; that is, which does not change the form of the ellipse.

Hence, in any given ellipse the parallelogram formed by tangents at the extremities of conjugate diameters, or the parallelogram on a pair of conjugate diameters taken in both length and direction, is constant.

Cor. If a and b denote the semiaxes, and a' and b' denote a pair of conjugate semidiameters, and θ be the angle between them,

$$a'b' \sin \theta = ab.$$

203. Let $APQB$ be a semicircle on a plane, U, and let CP and CQ be radii at right angles to one another. Let the whole be projected orthogonally upon a plane, V, passing through AB, and inclined to U at a fixed angle, θ. And let CP', CQ' be the projections of CP and CQ respectively.

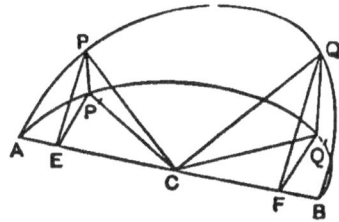

Then (Art. 201. Cor. 2), $AP'Q'B$ is a semiellipse, and CP' and CQ' are a pair of semi-conjugate diameters.

Draw PE and $QF \perp$ to AB, and join $P'E$ and $Q'F$. Then $P'E$ and $Q'F$ are $\perp$ to AB; and since PCQ is a right angle, $CF = EP$, and $EC = QF$. Also, $PP' = EP \sin \theta$, and $QQ' = FQ \sin \theta = CE \sin \theta$.

And

$$CP'^2 + CQ'^2 = CE^2 + EP^2 - PP'^2$$
$$+ CF^2 + FQ^2 - QQ'^2$$
$$= 2\,CP^2 - (EP^2 + FQ^2) \sin^2 \theta$$
$$= CP^2 (2 - \sin^2 \theta)$$
$$= \text{a constant.}$$

Therefore, the sum of the squares on a pair of conjugate diameters of an ellipse is constant.

Cor. Denoting the parts as in Art. 202. Cor.,

$$a'^2 + b'^2 = a^2 + b^2.$$

The results of Arts. 202 and 203 are known as the theorems of Apollonius.

204. Let V be a plane, cutting a circular cone so as to give an ellipse, and let P be any point on the ellipse. P is a point on the cone.

Let Z be the sphere which touches V at the focus S, and Z' be the sphere which touches V at the focus S'; also let K denote the circle of contact of Z with the cone, and K' denote the circle of contact of Z'.

Draw through P a generating line of the cone. This line cuts K in k, and K' in k', and at these points touches the spheres Z and Z'.

But K and K' being cone circles to the vertex O, kk' is constant for all positions of the generating line.

But $Pk = PS$ and $Pk' = PS'$, being tangents to the spheres.

$$\therefore SP + PS' = \text{a constant.}$$

Therefore, in any given ellipse, the sum of the distances of any point on the curve from the two foci is constant.

205. The result of the preceding article furnishes a convenient practical method of drawing an ellipse.

Over two pins placed at S and S' put a loop of inextensible thread, and keep it stretched by a pencil at P. The locus of P is an ellipse of which S and S' are the foci.

For the whole length of thread being constant, and the part SS' being constant, it follows that $SP + PS'$ is constant.

Cor. By considering the phase when P comes to A or to B, we readily see that $SB + BS' = AB$, and hence that the whole length of thread is $2AS'$, or $2BS$.

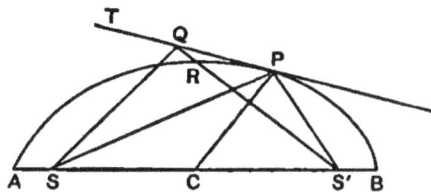

206. Let PT be a tangent to the ellipse at P, and let Q be any point, other than P, on this tangent. Then, since the tangent has only one point in common with the curve (Art. 193. Cor.), QS' cuts the curve in some point R.

Then, $\qquad SQ + QS'$ is $> SR + RS'$;

or $\qquad SQ + QS'$ is $> SP + PS'$.

Hence $SP + PS'$ is the shortest route from S to S' by way of the line PT, and hence SP and $S'P$ are equally inclined to PT.

Therefore, in any ellipse, the lines from the foci to the point of contact of any tangent are equally inclined to the tangent.

207. $O \cdot GQH$ is a circular cone; $APQB$ is an elliptic section, and CPD and EQF are circular sections cutting the elliptic section in the lines PM and QN.

CD and EF are diameters of the circles, and AB is the axis of the ellipse, all lying in the plane W (Art. 194). PM and QN are perpendicular to AB and to the lines CD and EF respectively.

Therefore $\quad CM \cdot MD = PM^2$, and $EN \cdot NF = QN^2$.

But $\triangle AMD \backsim \triangle ANF$, and $\triangle ENB \backsim \triangle CMB$.

$$\therefore AM : MD = AN : NF;$$

and $\qquad MB : CM = NB : EN$.

Therefore, by multiplication,

$$AM \cdot MB : MD \cdot CM = AN \cdot NB : NF \cdot EN;$$

or $\qquad AM \cdot MB : PM^2 = AN \cdot NB : QN^2$.

In a similar manner it is proved that a like relation holds true for the hyperbola. Therefore,

In the ellipse and the hyperbola, if perpendiculars be drawn from points on the figure to the principal axis, the squares on these perpendiculars are proportional to the rectangles on the parts into which each perpendicular divides the principal axis.

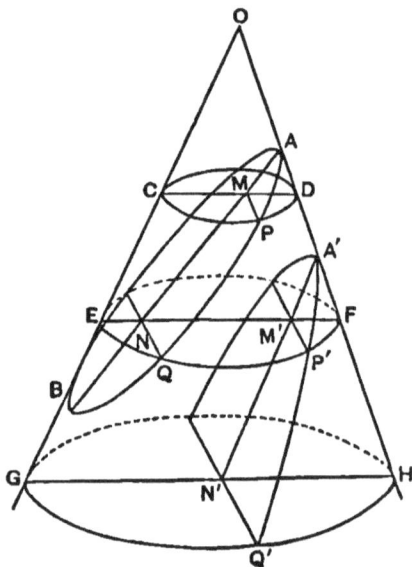

Again, let $A'P'N'$ be a parabolic section. Then $A'N'$ is parallel to OG, and $EM' = GN'$.

But from the similar $\triangle A'M'F$ and $A'N'H$,

$$A'M' : M'F = A'N' : N'H.$$

And $\qquad 1 : EM' = 1 : GN'$.

$$\therefore A'M' : EM' \cdot M'F = A'N' : GN' \cdot N'H;$$

or $\qquad A'M' : P'M'^2 = A'N' : Q'N'^2$.

Hence, in the parabola, the squares on perpendiculars from points on the figure to the axis are proportional to the parts of the axis intercepted between the vertex of the parabola and the foot of each perpendicular.

Def. The perpendiculars of this article are called *ordinates* to the principal axis, and the segments into which they divide the axis are called *abscissæ*. So that in the ellipse and hyperbola the square on any ordinate is in a constant ratio to the rectangle on its abscissæ.

In the parabola, however, one abscissa is infinite, and we have only the finite one to consider. Then, the square on an ordinate is in a constant ratio to its abscissa.

EXERCISES O.

1. Show that the area of an ellipse is πab.

2. A right cylinder has its base an ellipse with axes a and b. Show how to cut it by a plane so that the section may be a circle.

3. If P be a point on a hyperbola, of which S and F are foci, then $SP - PF =$ constant.

208. As the parabola is a limiting form of the ellipse (Art. 192. 2), the fundamental properties of the parabola may be obtained from those of the ellipse by supposing that one focus of the ellipse goes to infinity, while the other focus remains at a finite distance from the vertex.

We shall, however, obtain these relations by means of the perspective projection of the circle.

In the accompanying diagram, the right-hand figure is the projection of the left-hand one, and for the sake of convenience in comparison, a point and its projection are denoted by the same letter in both figures.

In (I), BB' is a tangent to the circle APB, touching it at B, and BAK is a centre line. $B'PQ$ is any secant line from B', and PT and QT are tangents meeting at T.

$B'V$ is a tangent from B' touching the circle at V, and TK is perpendicular to BA.

The circle is projected into a circular cone, and the plane which gives the parabola in section touches the circle at A and is parallel to the tangent BB', and BB' is projected to infinity.

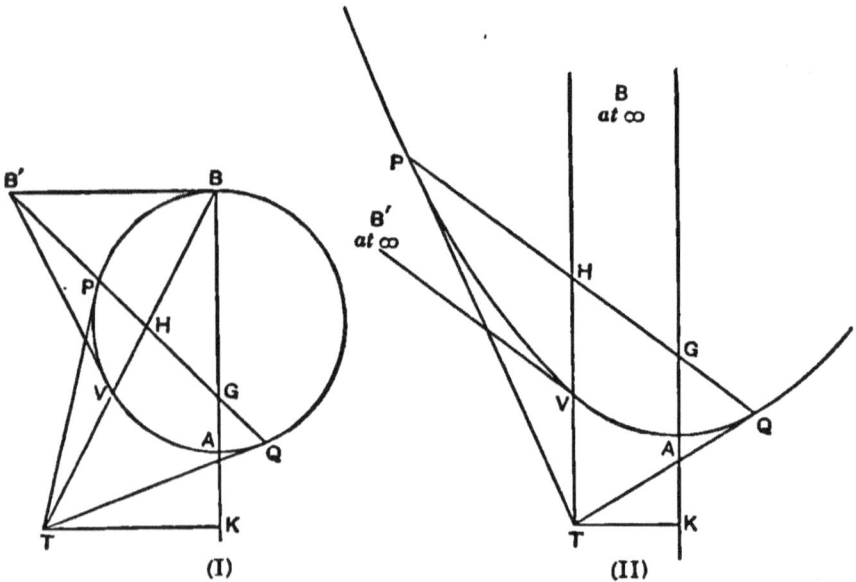

(I)   (II)

In the projection (II), A becomes the vertex A of the parabola, and BB' goes to infinity (Art. 181. Cor. 2), and hence the lines through B in (I) become a system of parallels in the projection (II), and thus in (II) $KAG\infty$ is the axis of the parabola, and $TVH\infty$ is parallel to the axis.

So, also, the lines through B' in (I) project into a system of parallels in (II); that is, QPB' and VB'

become the parallels $QP\infty$ and $V\infty$, or the tangent at V is parallel to PQ.

Now in (I) B' is pole to VB, and therefore $QHPB'$ is a harmonic range (P. Art. 311. 2).

Therefore, also, in (II), $QHP\infty$ is a harmonic range, and H is the middle point of QP.

Hence 1. In the parabola, any line parallel to the axis is a diameter, and bisects all chords parallel to the tangent ($V\infty$) at its vertex. Thus PQ is bisected by TVH.

The direction of the diameter conjugate to $VH\infty$ is given by PQ, but its position is at infinity.

Again, in (I), T is pole to PQ, and $TVHB$ is a harmonic range (P. Art. 311. 2).

Therefore, in (II), $TVH\infty$ is harmonic, and V bisects TH. Hence:

2. In the parabola, tangents at the end-points of any chord (PQ) meet upon the diameter to that chord (TH); and the part of the diameter intercepted between the chord and the point of meeting of the tangents is bisected by the curve. TH is bisected at V.

Again, in (I), TK is the polar of G (P. Art. 267), and hence $KAGB$ is a harmonic range.

Therefore, in (II), $KAG\infty$ is harmonic, and A bisects KG. Hence :

3. In the parabola, if two tangents be drawn from any point to the curve, and a perpendicular be drawn from the same point to the axis, the part of the axis intercepted between the foot of the perpendicular and the chord of contact (PQ) is bisected by the vertex of the curve. KG is bisected at A.

209. Since the circle can be projected into any conic, and anharmonic properties are projected without change, all the properties of the circle which depend upon anharmonic or harmonic relations are equally true for all the conics.

Thus the theorems in plane geometry, given in Arts. 311, 312, 313, 314, and many of the following ones, are true when we read conic for circle.

To enter any further into this subject is beyond the scope of this work. The student who desires to pursue this most interesting subject will find it fully developed in Salmon's 'Conics,' or Cremona's 'Projective Geometry,' or in Poncelet's great work, the 'Traité des propriétés projective des figures.'

EXERCISES P.

1. In the parabola, prove that the tangent at any point on the curve makes equal angles with the axis and the line joining that point to the focus.

2. P being a point on a parabola, if PM be drawn perpendicular to the axis and PN perpendicular to the tangent at P, the part MN intercepted on the axis is constant.

3. The tangent at the vertex of a parabola bisects the part of any other tangent lying between the point of contact and the axis.

4. The tangent at the vertex of a parabola, and the perpendicular from the focus upon any other tangent, meet the latter tangent at the same point.

5. In the projection of Art. 208, if TK (1) projects into the directrix, then G will project into the focus.

SECTION 3.

SPHERIC GEOMETRY.

210. When a spatial figure is cut by a sphere, the elements common to the spatial figure and the sphere form a figure which lies on the sphere in the same manner as a plane figure lies upon its plane.

Such a figure is a *spheric* figure, as being confined to a spherical surface, and the geometry of such figures is called *spheric geometry* or *spherical surface geometry*.

On account of the uniform curvature of a sphere, a spheric figure may be moved about upon the spherical surface upon which it lies, without undergoing any necessary change in the relations of its parts, just as a plane figure may be moved about over the plane surface upon which it lies.

There are many analogies between spheric geometry and plane geometry, and many of the theorems and of the methods employed are more or less alike. But there are also fundamental differences.

The prominent analogies will be exhibited in the sequel.

211. It must be understood in the beginning that spheric geometry does not deal with a comparison of the properties or relations of figures drawn upon different spheres; its purpose is not this, but to investigate the properties which belong to a figure *in consequence of its lying upon a sphere.*

Hence all spheric figures are supposed to lie on one and the same sphere, just as all the figures in plane geometry are supposed to lie on one and the same plane.

The radius of the particular sphere is altogether arbitrary, and, except in the case of metrical theorems or problems, the radius may be left out of the consideration.

The centre of the sphere will be referred to as *the centre.*

212. Every section of a sphere by a plane is a circle, and when the plane contains the centre of the sphere, the section is the largest circle in this way obtainable, and is called a *great* circle of the sphere.

It will appear hereafter that when a plane figure involving the line has an analogue in spheric geometry, the line is represented by the great circle. And as there can be no straight line in connection with any spheric figure, we shall, for the sake of the analogy, commonly speak of a great circle as a *spheric line.* Then all other circles are spheric circles.

Any limited part of a spheric line is a *spheric arc,* and parts of other circles are *circular arcs.*

213. The spheric line, unlike a planar line, returns into itself without passing to infinity.

Evidently the spheric line divides the whole spherical surface into two congruent parts, just as the planar line may be said to divide the whole plane into two congruent parts. The parts of the plane, however, extend to infinity, while those of the spherical surface do not.

We have here a fundamental distinction between plane and spheric geometry, namely, that *spheric geometry has no infinity.*

214. As any plane which gives in section a spheric line must pass through the centre, any two points on the sphere, not in line with the centre, determine, with the centre, one plane, and therefore one spheric line.

Thus, like a planar line, a spheric line is determined by any two points, provided they are not collinear with the centre.

And through any two points not collinear with the centre, one, and only one, spheric line can pass.

When two points are in line with the centre, the three points determine only one line, a diameter of the sphere, and through this any number of planes can pass, giving in section the same number of spheric lines.

Now in plane geometry, any two points determine one line, unless the points be at infinity. For in this latter case, since all parallels of a system meet at infinity, two points at infinity, in opposite directions, determine a system of parallels.

We see, then, that for two points to be collinear with the centre, in spheric geometry, is analogous to two points at infinity in opposite directions, in plane geometry.

Thus spheric lines passing through a pair of opposite points are analogous to parallel lines in plane geometry; and hence *there is no theory of parallels in spheric geometry* as in plane geometry.

Cor. Any number of points on a sphere, no two of which are collinear with the centre, and no three of

which are complanar with the centre, determine as many
spheric lines as there are groups of the points taken two
and two.

The corresponding theorem in plane geometry is, that
any number of points, no two of which are at infinity,
and no three of which are in line, determine as many
lines as there are groups of the points taken two and two.

215. The normal, through the centre, to the plane of
a great circle, meets the sphere in two opposite points
which are end-points of a diam-
eter.

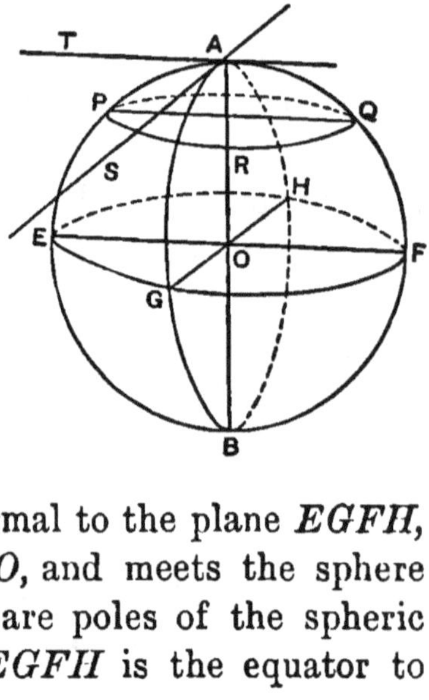

These points are *poles* of the
great circle; and in relation to
the poles the circle is called the
equator.

Thus every spheric line has
two poles, and any point on the
sphere, considered as a pole,
has an opposite pole, and an
equator.

Thus *AB*, in the figure, is normal to the plane *EGFH*,
and passes through the centre *O*, and meets the sphere
at *A* and at *B*. Then *A* and *B* are poles of the spheric
line *EGFH*, and reciprocally *EGFH* is the equator to
the poles *A* and *B*.

Evidently the angle *AOE*, subtended at the centre
between a pole and any point on its equator, is a right
angle, and the spheric arc *AE* is one-fourth of a whole
circumference.

If a quadrant of a great circle has one extremity fixed
at *A* while the other moves over the surface of the

sphere, the moving extremity will describe the great circle or spheric line *EGFH*, which is the equator to the point *A* as pole.

Hence the quadrant *AE* is the *spheric radius* of the great circle described; or the spheric radius of a great circle is a quadrant.

Cor. If any point *P* be taken in the quadrant *AE*, when the quadrant moves over the surface having *A* fixed, *P* will describe a circle *PRQ* upon the sphere. As the arc *AP*, and therefore the chord *AP*, is constant, *PRQ* is a cone-circle to *A* as vertex, and its plane is normal to *AB*, and therefore parallel to the plane of *EGFH*.

And as *A* and *B* are points on the sphere from which the spheric circle *PRQ* can be described by means of constant arcs, *AP* or *BP*, these points are poles to the circle *PRQ*, and the arc *AP*, as also *BP*, is its spheric radius.

Thus every circle on a sphere has two spheric radii which are supplementary to one another, and the poles of any great circle are poles to all spheric circles whose planes are parallel to that of the great circle.

SECTION OF TWO PLANES.

216. The sphere which has its centre on the common line of two intersecting planes has, in section by these planes, two spheric lines which intersect in opposite points, or at the end-points of a diameter.

Thus the two planes which have in common the line *AB* (Fig. of 215) give by their intersection with the

sphere the two spheric lines $AEBF$ and $AGBH$, intersecting in A and B, and mutually bisecting one another.

The angle between these spheric lines is equal in
measure to the dihedral angle between the planes which
give rise to the lines. But if $EGFH$ be the equator to
A, EO and GO are each perpendicular to AB, and therefore EOG measures the dihedral angle between the
planes, and hence also the angle between the spheric
lines.

But the angle $EOG = (\text{arc } EG) \div EO$; and since EO
is supposed to be constant, our investigations being confined to one and the same sphere,

Therefore, the angle between two spheric lines is
proportional to the arc which they intercept upon the
equator to their points of intersection as poles.

Cor. 1. If, through A, tangents to the spheric lines
be drawn, AT to $AEBF$ and in its plane, and AS to
$AGBH$ and in its plane, the angle TAS is equal to EOG,
and is the angle between the spheric lines.

Therefore, the angle between two tangents drawn to
two spheric lines at their point of intersection is the
angle between the spheric lines.

Cor. 2. When the angle EOG is a right angle, OG is
normal to the plane of $AEBF$, and G is a pole to the
circle $AEBF$.

Therefore, two spheric lines are perpendicular to one
another when one of them passes through a pole of the
other; and in this case each passes through both poles of
the other.

217. The spheric figure $AEBGA$, formed by two
spheric lines between their points of intersection, is a

lune. The points *A* and *B* are vertices of the lune, and the angle between the spheric lines forming its sides is the angle of the lune.

Evidently every lune is accompanied by an opposite congruent lune, as *AEBGA* and *AFBHA*; and any two spheric lines divide the whole spheric surface into four lunes which are congruent in opposite pairs.

We have the analogous case in plane geometry, where any two intersecting lines divide the whole plane into four sections, which, although extending to infinity, may properly be said to be congruent in opposite pairs.

It will be seen from this and other cases that the analogy between plane and spheric geometry is descriptive rather than metrical in kind.

THREE PLANES — SECTION OF THREE-FACED CORNER OR TRIHEDRAL ANGLE — SPHERIC TRIANGLE.

218. Just as a plane section of any corner is a plane polygon with a side corresponding to and given by each face of the corner, so the section of a corner by a sphere with its centre at the vertex of the corner is a spheric polygon, whose sides are parts of spheric lines, and the number of whose sides is the same as that of the faces forming the corner.

In spheric as in plane geometry the most important polygon is the *triangle.*

In plane geometry three given lines can form but one triangle, since they determine at most but three points. But, as every spheric line meets every other spheric line in two points, three spheric lines, which are not concurrent, determine six points, and these may combine in threes to form eight spheric triangles.

Therefore, any three non-concurrent spheric lines divide the surface of the sphere into eight triangles.

219. Let $AHA'G$, $BIB'J$, and $CFC'E$ be three non-concurrent spheric lines. They meet in the six points A, B, C, A', B', C', of which A is opposite A', B opposite B', and C opposite C'.

The eight determined triangles are

$$ABC, \ ABC', \ AB'C, \ A'BC,$$

$$A'B'C', \ A'B'C, \ A'BC', \ AB'C'.$$

Since A is opposite A', etc., the arc $AB =$ the arc $A'B'$.

Similarly, arc $BC =$ arc $B'C'$ and arc $CA =$ arc $C''A'$.

Also, as $ACC'A'$ and $ABA'B'$ determine two planes, the angle at A is equal to the angle at A'; and so also the angle at B is equal to the angle at B', and the angle at C to the angle at C'. And thus the opposite triangles ABC and $A'B'C'$ have all their parts in the one respectively equal to the corresponding parts in the other. But the triangles are not superposable. For taking the centre as a point of reference, ABC and $A'B'C'$ are in opposite orders of rotation.

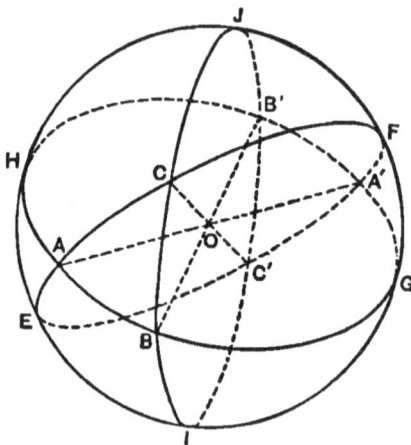

In the case of plane triangles we could invert one of them, or turn it over in the plane, and then superimpose them; but this operation is clearly impossible in a spheric figure. Spheric triangles related in this manner are *symmetrical*, or *conjugate*, to one another, and they are evidently produced by the sections of two symmetrical three-faced corners having a common vertex at the centre (Art. 39).

The eight triangles are symmetrical or conjugate in opposite pairs as follows:

ABC and $A'B'C'$, ABC' and $A'B'C$, $AB'C$ and $A'BC'$, $A'BC$ and $AB'C'$.

220. Let ABC (Fig. of 219) be a spheric triangle formed by section of the three-faced corner $O \cdot ABC$.

The angles at A, B, and C, whose measures are respectively those of the three dihedral angles of the corner, are called the *angles* of the triangle, and are usually denoted by A, B, C; and the arcs BC, CA, and AB are called the sides of the triangle, and are denoted by a, b, c.

Here two views confront us.

If linear units are to be introduced, and arcs are to be considered with respect to length, the length of the radius of the sphere is involved, and our investigations are confined to some one sphere whose radius is known or determinable with reference to the unit of measure employed.

On the other hand, if our operations and results are to have no reference to the length of the radius, we must take for the side of the spheric triangle, not the arc itself, but its ratio to the radius.

This ratio is an angle, a face angle of the corner which in section gives the spheric triangle.

Spheric geometry then ceases to have any distinct relation to the sphere, except in name, involves no relation of length, and becomes a geometry of direction only.

In what follows we shall not confine ourselves exclusively to either view, but shall adopt that which serves our purpose at the time.

In the majority of applications, however, the second one is the only view that can be adopted.

Thus in applying the results of spheric geometry to the visible surface of the heavens, anything like a linear unit is out of the question.

According to the second view, a spheric triangle consists of six parts, all angles.

Three alternate parts, called the angles of the triangle, are respectively equal in measure to the dihedral angles of a three-faced corner, and the remaining three, called the sides of the triangle, are respectively equal to the face angles of the same corner.

And thus all the relations which exist between the dihedral angles and face angles of a three-faced corner, exist also between the angles and sides of a spheric triangle.

The adaptation to a triangle described on a given sphere is easily effected; for it is only necessary to express the sides in radians and then multiply the result by the radius of the sphere.

221. In any three-faced corner the sum of two face angles is greater than the third (Art. 35).

Therefore, the sum of two sides of a spheric triangle is greater than the third side; and the difference between any two sides of a spheric triangle is less than the third side of the triangle.

222. Theorem. The shortest path from one point to another, on the surface of a sphere, is along the spheric line through the points.

Let A and B be the points, and C be any point on the spheric join AB. With A and B as poles describe circles PCD and QCE to pass through C. Take P, any point on the circle PCD, and draw the spheric arcs AP, and PB cutting the circle QCE in Q.

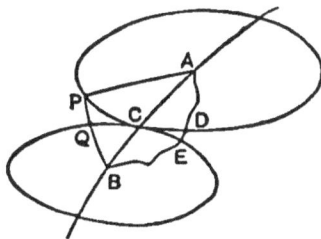

Then APB is a spheric triangle, and

$$AP + PB > AB. \tag{221}$$

Therefore, P lies without the circle QCE, and the circles PCD and QCE touch at C.

Now let $ADEB$ be any path on the sphere from A to B. Then the path from A to D may be brought to extend from A to C by turning the circle PCD about its centre A, until D comes to C. In a similar way, the path from B to E may be brought to extend from B to C.

But by this change the whole path is shortened by the distance DE. That is, the path is shortened by making it pass through C, a point on the spheric line through A and B.

In like manner, each part of the path is shortened by making it pass through some arbitrary point on the spheric line AB.

Hence the path is shortest when every point lies on the spheric line from A to B.

Cor. 1. When two spheric circles touch, the spheric line through their poles passes through their point of contact.

Cor. 2. When two points on a sphere are in line with the centre, an indefinite number of equal shortest paths may be drawn from one point to the other.

223. In any three-faced corner, considering two face angles and the opposite dihedral angles, the greater face angle is opposite the greater dihedral angle; and conversely, the greater dihedral angle is opposite the greater face angle (Art. 40).

Hence in any spheric triangle, considering two sides and the opposite angles, the greater side is opposite the greater angle; and conversely, the greater angle is opposite the greater side.

Cor. 1. If a spheric triangle has two equal sides, it has two equal angles; and conversely, if it has two equal angles, it has two equal sides.

Cor. 2. The order of magnitude of A, B, C, the angles of a spheric triangle, is the same as that of a, b, c, the sides of the triangle.

Hence spheric triangles, like plane ones, are equilateral, and isosceles, and scalene.

224. When the three face angles of a three-faced corner are given, the dihedral angles also are given (Art. 41. Cor. 2); and conversely, when the dihedral angles are given, the face angle also are given (Art. 44. Cor. 2). Therefore,

1. When the sides of a spheric triangle are given, the angles also are given, and all the parts are known.

This case holds also for plane triangles.

2. When the three angles of a spheric triangle are given, the sides also are given, and all the parts are known.

This does not hold for plane triangles, and this fundamental difference between spheric and plane geometry is due to the fact that spheric geometry has no theory of similar figures, a theory which plays so important a part in plane geometry. Similarity requires an equality of tensors, and therefore involves the consideration of linear extension. But in a spheric triangle, where all the parts are angles, there is no place for linear extension, and hence no similarity exists beyond absolute equality.

Similar spheric triangles might be drawn upon spheres of different radii, but the comparison of these, although belonging to spatial geometry, does not belong to spheric geometry (Art. 211).

225. In any corner the sum of the face angles is less than a circumangle (Art. 42).

Hence the sum of the sides of a spheric triangle is less than a circumangle; or if we introduce the radius, is less than a circumference.

Cor. 1. When two sides of a spheric triangle become straight angles, the third side vanishes and the figure becomes a lune.

Cor. 2. When each side becomes a right angle, the

planes forming the faces of the corner become the rectangular co-ordinate planes of space (Art. 8. Cor.), and each angle becomes a right angle.

Hence a spheric triangle which has each side a right angle has also each angle a right angle. Such a triangle is a *quadrantal* triangle.

Cor. 3. If two sides of a spheric triangle be produced to meet, a second triangle is formed which is said to be *co-lunar* with the first. These two triangles have an angle and the opposite side respectively equal, while the remaining two sides in the one are supplementary to the corresponding sides in the other, and the remaining angles in the one are supplementary to the remaining angles in the other.

Cor. 4. Any spheric triangle has three colunar triangles.

226. The polar triangle. When the vertices of one spheric triangle are poles to the sides of another spheric triangle, the first triangle is said to be *polar* to the second. And the second triangle is also polar to the first.

Let A', B', C' be poles of a, b, c respectively, and let a', b', c' be the sides of the spheric triangle having A', B', C' as vertices.

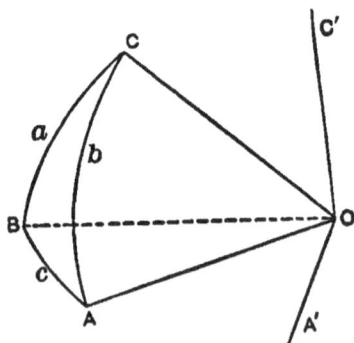

Since C' is pole to AB, $C'O$ is $\perp$ to OB (Art. 215); and since A' is pole to BC, $A'O$ is $\perp$ to OB.

Therefore, OB is normal to the plane of OA' and OC',

and B is therefore pole to the spheric join of $A'C'$; that is, to b'.

Similarly, A is pole to a' and C to c'; and the spheric triangle ABC is polar to $A'B'C'$.

Hence when one spheric triangle has its vertices poles to the sides of a second triangle, the vertices of the second triangle are poles to the sides of the first, and the two triangles are polar to each other.

Cor. 1.   A quadrantal triangle is polar to itself.

Cor. 2.   The points A', B', C' determine eight triangles (Art. 218), every one of which might be said to be polar to ABC.   Or more generally, A, B, C determine one set of eight triangles, and A', B', C' determine a second set; and every triangle of one set has every triangle of the other set as a polar triangle.

It is easily seen, however, that two triangles in either set are conjugate (Art. 219), and the remaining six are co-lunars of these.   So that if we agree that the spheric triangle formed from three given spheric lines is to be considered as being the triangle (triangle or its conjugate), each of whose sides are less than a straight angle or a semi-circumference, then each spheric triangle has but *one* polar triangle (a triangle or its conjugate).

227. The spheric triangles ABC and $A'B'C'$ being polar to one another, $A'O$ is normal to the plane of a, and $B'O$ is normal to the plane of b.   But the angle between two planes is the supplement of the angle between normals to the planes (Art. 16. Def. 2); and the angle between the planes is the angle C, and the angle between the normals is the side c'.

$$\therefore \ C + c' = \pi;$$

Or the sum of an angle of a spheric triangle and the corresponding opposite side of the polar triangle is a straight angle.

Cor. $A+a'=B+b'=C+c'=A'+a=B'+b=C''+c=\pi.$

228. From the preceding article,

$$A + B + C + a' + b' + c' = 3\pi.$$

But (Art. 225) $a'+b'+c'>0$ and $<2\pi$;

$$\therefore A+B+C \text{ is } <3\pi \text{ and } >\pi.$$

That is, the sum of the angles of a spheric triangle is variable, lying between the limits of two right angles and six right angles. And hence, in every spheric triangle the sum of the angles exceeds two right angles.

The amount by which the sum of the three angles exceeds a straight angle is called the *spherical excess* of the triangle. If we denote it by *E*, we have

$$E = A + B + C - \pi.$$

229. It has been shown (Art. 219) that conjugate spheric triangles, although having all their corresponding parts respectively equal, are not superposable, but correspond to one another after the manner of the right and the left hand.

Hence in the determination of a spheric triangle from its parts, there is always the kind of ambiguity which results from not knowing whether a particular triangle or its conjugate is the one required.

This ambiguity disappears in the case of an isosceles spheric triangle, for this triangle is conjugate to itself.

For the sake of simplicity, then, we shall agree in what follows that a spheric triangle is given or determined when we know the triangle or its conjugate.

230. A spheric triangle is given when the three sides are given, or when the three angles are given.

This follows from Art. 41. Cor. 2, and Art. 224.

231. A spheric triangle is given when two sides and the included angle are given.

Let ABC, $A'B'C'$ be two spheric triangles in which $\angle C' = \angle C$, $a' = a$, and $b' = b$; and let the parts be disposed in the same order.

Place C' on C and $C'A'$ along CA. Since $b' = b$, A' coincides with A. Also, since $\angle C' = \angle C$, the side a' will lie along the side a, and as $a' = a$, B' will coincide with B.

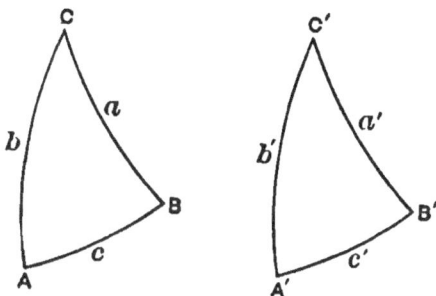

And as A and B determine only one spheric line, $A'B'$ coincides with AB, and the triangles coincide in all their parts.

Therefore, the triangle ABC is given when two sides and the included angle are given (see P. Art. 66).

232. A spheric triangle is given when two angles and the included side are given.

Let A and B and the side c be given.

Then the sides a' and b' and the included angle C' is given for the polar triangle, and therefore the polar triangle is given (Art. 231).

Hence the original triangle is given.

233. Let ABC be an isosceles spheric triangle with $CA = CB$, and hence the $\angle A = \angle B$.

Draw the spheric line $D'CD$ to the middle point, D, of the base AB. Then CD is median to the base.

The triangles ACD and BCD having AC and AD respectively equal to BC and BD, and the included angles A and B equal, are conjugate.

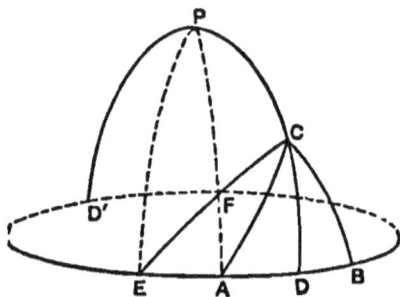

Therefore, $\angle CDA = \angle CDB = \neg,$

and $\angle ACD = \angle BCD.$

Hence the median to the base of an isosceles spheric triangle is the right bisector of the base, and the bisector of the vertical angle.

Hence, also, every point on the spheric line $D'CD$ is equidistant from A and B, distance being measured along a spheric line.

In the same manner as in plane geometry (P. Art. 54) it is shown that every point equidistant from A and B, and lying on the spheric surface, is on the right bisector of the base AB, i.e. on the spheric line $D'CD$.

Cor. In spheric geometry the join of A and B is either ADB or $AD'B$; i.e. there are two joins whose sum makes up the whole spheric line.

The bisector of one of these joins evidently bisects the other also; as at D and D'.

234. Let P (Fig. of 233) be the pole of AB, and let C be any point, lying between P and the arc AB, on the spheric line PD.

Then $PA = PD$, and PA is $< PC + CA$, (Art. 221.)

and $\therefore$ CD is $< CA$.

And $\angle PAD = \angle PDA = \daleth$, and $\angle CAD$ is $< \daleth$.

Therefore, the least distance from C to the spheric line AB is along the perpendicular CD.

Also, two equal spheric lines, CA and CB, can be drawn from C to AB, and these lie upon opposite sides of CD, and are equally inclined to it, and meet the line AB at equal distances from the foot of the perpendicular.

235. From C draw any spheric line, CE, to meet AB, and to cut PA at some point F between P and A.

Then, PE is $< PF + FE$, (Art. 221.)

and CA is $< CF + FA$;

Therefore, adding, $PE + CA < CE + PA$.

But $PE = PA$;

$\therefore$ CA is $< CE$.

This holds true so long as CE intersects PA between P and A. But as E recedes from A along the spheric line AB, CE will continue to meet PA between P and A until E comes to D', opposite D.

Then if E be supposed to start from D and make a complete circuit along the spheric line AB, CE, starting from the value CD, will increase as DE increases, until

E comes to *D′*, when it has its maximum value. It then decreases until *E* returns through *B* to *D*, at which position *CE* has its minimum value.

Cor. 1.   Since all spheric lines are of the same length, or contain the same angle, the greatest and least spheric arcs from a point on the sphere to any spheric line, *AB*, are the parts of the spheric line perpendicular to *AB*, which are intercepted between *C* and *AB*.

Cor. 2.   If two equal spheric arcs be drawn from a point on the sphere to a spheric line which is not its equator, they are equally inclined to the longest spheric arc from the given point to the given line, and lie upon opposite sides of it.

236.   As in a plane triangle, so in a spheric one, when two sides and an angle opposite one of them are given, the triangle may be ambiguous. Owing to the facts, however, that any two spheric lines intersect in two points, and that the sum of the angles of a spheric triangle is not a fixed quantity, the condition for ambiguity is much more complex than in plane geometry.

Also, unlike a plane triangle, a spheric triangle may be ambiguous when two angles and a side opposite one of them are given.

THE AMBIGUOUS CASE.

237.   In the following examination we assume that the relative magnitudes of the parts given are such as to determine a *real* triangle, so that we shall not be concerned with conditions which lead to impossible or vanishing triangles, although such conditions are easily obtained.

For the given parts let us take the sides a and b and the angle A, and let us consider the subject under three cases, according as A is less, equal to, or greater than, a right angle.

For a general diagram let ACD and AED be two spheric lines at right angles, and let C be a pole of AED. Through A and D draw the spheric line APD, making the angle CAP less than a right angle, and the spheric line AQD, making the angle CAQ greater than a right angle.

Take any point C' between A and C, and any point C'' between C and D.

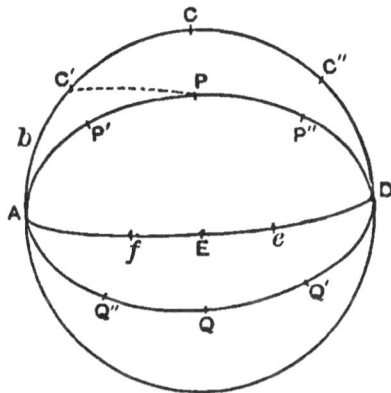

Let the side b be measured from A along the arc ACD, and let the given angle A be the angle at A. Then a is drawn from some point on the arc ACD to the arc APD or AED or AQD, according as A is less than a right angle, equal to a right angle, or greater than a right angle. One such triangle is represented at $AC'P$, where a denotes the side $C'P$.

CASE I. Let $A = \angle CAP, < \frac{\pi}{2}$.

Then the shortest distance from a point on ACD to the arc APD is perpendicular to APD (Art. 234), and is therefore not along ACD.

1. Let $b = AC'$, $< \frac{\pi}{2}$, and let P' be the foot of the perpendicular from C'.

Two equal arcs can be drawn from C' to APD, one on each side of $C'P'$ (Art. 234), and the triangle may be ambiguous.

For a case of ambiguity, however, a must be $< C'A$, i.e. $< b$.

$\therefore A < \frac{\pi}{2}$, $b < \frac{\pi}{2}$, $a < b$ is ambiguous.

2. Let $b = AC = \frac{\pi}{2}$, and let P be the foot of the perpendicular from C.

Then evidently the case is ambiguous if a is less than AC and greater than CP, and CP, being on the equator to A, measures the angle A (Art. 216).

$\therefore A < \frac{\pi}{2}$, $b = \frac{\pi}{2}$, $a \begin{smallmatrix} > A \\ < b \end{smallmatrix}$ is ambiguous.

3. Let $b = AC''$, $> \frac{\pi}{2}$. Then the case will be ambiguous if $a < C''D$.

$\therefore A < \frac{\pi}{2}$, $b > \frac{\pi}{2}$, $a < (\pi - b)$ is ambiguous.

Cor. In all the foregoing cases it is readily seen that the ambiguity disappears when the triangle becomes right angled by a being drawn perpendicular to the arc APD.

CASE II. Let A be the angle $CAE = \frac{\pi}{2}$.

Since C is a pole of AED, from any point on ACD, two equal spheric arcs can be drawn to AED, one lying on each side of AC and equally inclined to it (Art. 234). The two triangles thus formed would be conjugate and not ambiguous (Art. 229). But if the point C be taken, all spheric arcs from C to AED are equal, and the triangle is indeterminate.

Therefore, with $A = \frac{\pi}{2}$ there is no real ambiguity, but when $b < \frac{\pi}{2}$ and $a > b$, and also when $b > \frac{\pi}{2}$ and

$a > (\pi - b)$, two triangles are obtained which are conjugate; and when $b = \frac{\pi}{2}$, the triangle is indeterminate.

CASE III. Let A be the angle $CAQ > \frac{\pi}{2}$. Since the angle CAQ is greater than a right angle, the longest spheric arc from any point on ACD to AQD is perpendicular to AQD, and is therefore not along ACD.

1. Let $b = AC'$, $< \frac{\pi}{2}$; and let Q' be the foot of the perpendicular from C' to AQD.

Then two equal spheric segments may be drawn from C' to the arc AQD, one on each side of $C'Q'$ (Art. 235. Cor. 2), and the triangle may be ambiguous. For a case of ambiguity, however, $C'Q'$ must be greater than $C'D$, or $a > (\pi - b)$.

$\therefore A > \frac{\pi}{2}$, $b < \frac{\pi}{2}$, $a > (\pi - b)$ is ambiguous.

2. Let $b = AC = \frac{\pi}{2}$; then if Q be the foot of the perpendicular from C, it is readily seen that the case will be ambiguous if a is less than CQ and greater than CD. But CQ being on the equator to A measures the angle A (216).

$\therefore A > \frac{\pi}{2}$, $b = \frac{\pi}{2}$, $a \begin{smallmatrix} < A \\ > b \end{smallmatrix}$ is ambiguous.

3. $b = AC''$, $> \frac{\pi}{2}$. Then, if Q'' be the foot of the perpendicular from C'', the triangle will be ambiguous if a lies between $C''Q''$ and $C''A$.

$\therefore A > \frac{\pi}{2}$, $b > \frac{\pi}{2}$, $a > b$ is ambiguous.

238. The results of the preceding article are collected in the following table:

	$b < \frac{\pi}{2}$	$b = \frac{\pi}{2}$	$b < \frac{\pi}{2}$
$A < \frac{\pi}{2}$	$a < b$ ambiguity.	$a > A < b$ ambiguity.	$a < (\pi - b)$ ambiguity.
$A = \frac{\pi}{2}$	$a > b < (\pi - b)$ conjugate.	$a = b$ indeterminate.	$a < b > (\pi - b)$ conjugate.
$A > \frac{\pi}{2}$	$a > (\pi - b)$ ambiguity.	$a < A > b$ ambiguity.	$a > b$ ambiguity.

We see from the table that ambiguity occurs only when A is not a right angle.

239. By making use of the polar triangle we can readily investigate the cases of ambiguity when two angles and a side opposite one of them are given. For when a triangle is ambiguous, its polar is ambiguous, and *vice versa.*

The table corresponding to that of the last article is here given:

	$B > \frac{\pi}{2}$	$B = \frac{\pi}{2}$	$B < \frac{\pi}{2}$
$a > \frac{\pi}{2}$	$A > B$ ambiguity.	$A < a > B$ ambiguity.	$A > (\pi - B)$ ambiguity.
$a = \frac{\pi}{2}$	$A < B > (\pi - B)$ conjugate.	$A = B$ indeterminate.	$A > B < (\pi - B)$ conjugate.
$a < \frac{\pi}{2}$	$A < (\pi - B)$ ambiguity.	$A > a < B$ ambiguity.	$A < B$ ambiguity.

240. Theorem. In any right-angled spheric triangle the number of sides which are less than, equal to, or greater than, a right angle is at least two.

In the figure of Art. 237, let A be a right angle.

1. If $b = \frac{1}{2}$, then $a = \frac{1}{2}$, and two sides are right angles.

2. Let $b = AC'$, and let E be the pole of ACD. Then, $C'E = \frac{1}{2}$, and $C'f$ is $< \frac{1}{2}$, and $C'e$ is $> \frac{1}{2}$. But, also, Af is $< \frac{1}{2}$, and Ae is $> \frac{1}{2}$.

And a and c are both less, or equal to, or greater than, $\frac{1}{2}$.

3. Let $b = AC''$. Then $b > \frac{1}{2}$; and $c''e$ is $< \frac{1}{2}$, and $C''f$ is $> \frac{1}{2}$. But Ae is $> \frac{1}{2}$, and Af is $< \frac{1}{2}$.

Therefore, when $b < \frac{1}{2}$, both a and c are less than, equal to, or greater than, $\frac{1}{2}$.

When $b = \frac{1}{2}$, both a and b are equal to $\frac{1}{2}$.

When $b > \frac{1}{2}$, b and c are $> \frac{1}{2}$, or a and c are $= \frac{1}{2}$, or b and a are $> \frac{1}{2}$.

And the theorem is proved.

241. Theorem. The sum of two angles of a spheric triangle, and the sum of the sides opposite these angles, are both less than, equal to, or greater than, a straight angle.

Let CDF be a lune, and let EF be equator to C and D.

Through G, the middle point of EF, draw the spheric line AB, meeting the sides of the lune in A and B. Then CAB is a spheric triangle.

It is evident that the triangles CAB and DBA are congruent, and therefore that $\angle CAB = \angle DBA$, and $\angle CBA = \angle DAB$, etc.

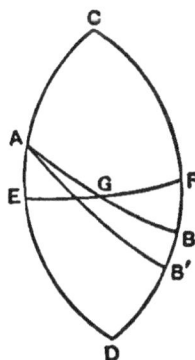

Then $\angle CAB + \angle CBA = \angle CAB + \angle BAD = \pi$,

and $\qquad CA + CB = CA + AD = \pi$.

Therefore, if the sum of two angles is a straight angle, so also is the sum of the two opposite sides.

Now take B', any point between B and D, and draw the spheric arc AB'.

Then

$$\angle ABB' + \angle BAB' + \angle BB'A > \pi; \quad \text{(Art. 43. Cor. 2.)}$$

and $\qquad \angle ABB' = \angle CAB.$

$$\therefore \ \angle CAB' + \angle BB'A > \pi.$$

Also, $\qquad CA + CB' > CA + CB > \pi.$

Therefore, when the sum of two angles is greater than a straight angle, the sum of the opposite sides is also greater than a straight angle.

And since these sums decrease and increase together, the theorem follows.

Hence $\frac{1}{2}(A + B)$ and $\frac{1}{2}(a + b)$ are both $> \frac{\pi}{2}$, both $= \frac{\pi}{2}$, or both $< \frac{\pi}{2}$.

This relation is commonly expressed by saying that $\frac{1}{2}(A + B)$ and $\frac{1}{2}(a + b)$ are of the same *affection*.

EXERCISES Q.

1. The area of a spheric triangle is Er^2, where E is the spherical excess.

2. The area of a spheric polygon is $\{\Sigma A - (n - 2)\pi\}r^2$, where ΣA is the sum of the angles, and n is the number of sides.

3. The area of an equilateral spheric triangle is one-fourth that of the surface of the sphere. Show that its angle is 120°, and find its side.

4. AB is a spheric arc, and C is its middle point. The locus of P, such that $\angle APC = \angle BPC$, is two spheric lines perpendicular to one another. Prove this, and state its analogue in plane geometry.

5. If the direction from A to B, two places on the earth, is estimated along a spheric line, and in terms of the angle which this line makes with the meridian of the first place, show that if A and B have different latitudes and longitudes the direction from A to B is not the opposite of the direction from B to A.

6. If a spheric triangle be formed by cutting a three-faced corner by a sphere, the centre of the sphere being the vertex of the corner, show (i.) that the isoclinal line to the edges of the corner gives the centre of the circle circumscribing the spherical triangle ; (ii.) That the isoclinal line to the faces gives the centre of the inscribed circle of the triangle.

7. What are given by the external isoclinal lines to the corner?

8. A spheric line is described by a quadrant which has one extremity fixed (215) ; what is the analogue in plane geometry?

MISCELLANEOUS EXERCISES.

1. Non-parallel lines do not necessarily intersect.

2. Two circles in space may pass within one another and have two, one, or no points in common.

3. From the definition of a tangent (P. Art. 109) show that the tangent to a circle lies in the plane of the circle.

4. A plane which is normal to the common line of two planes is perpendicular to both planes.

5. If any number of planes meet in parallel lines, the normals to these planes, from the same point, are complanar.

6. The sum of the normals from a point A to the planes U and V is the same as that of the normals from B to the same planes. Show that if P be any point in the line AB, the sum of the normals from P to U and V is constant.

7. Show that Ex. 6 holds good for any number of planes, U, V, W, etc.

8. If the sum of the normals to the planes U and V be the same for any three points, A, B and C, it is the same for every point in the plane of ABC.

9. The right-bisector plane of the common perpendicular to two lines bisects the join of any two points, one on each line.

10. A perpendicular is drawn to the base of a regular pyramid and meets the faces, produced where necessary. Then, the sum of the distances of the points of intersection from the base is constant.

11. Find in a given plane, a point equidistant from three given points.

12. Determine on a given line the point which is equidistant from any two given points.

13. The bisecting plane of a dihedral angle of a tetrahedron divides the opposite edge into segments which are proportional to the areas of adjacent faces.

14. The shortest chord through any point within a sphere is normal to the diametral plane containing the point.

15. If one intersection of a sphere by a cylinder is a circle, so also is the other intersection.

16. If one intersection of a sphere by a cone is a circle, so also is the other intersection.

17. The perpendiculars to the faces of a tetrahedron, at their centroids, are concurrent.

18. The vertices of a cuboid are conspheric.

19. The edges of a given three-faced corner pass through three fixed points. Show that its vertex is fixed.

20. The medians of a tetrahedron, taken in both length and direction, form a quadrilateral.

21. In a three-faced corner the planes through the edges bisecting the dihedral angles form an axial pencil.

22. In a three-faced corner the planes through the bisectors of the face-angles and perpendicular to the faces form an axial pencil. How many such pencils in all?

23. What is the locus of a point equidistant from two given points?

24. What is the locus of a point equidistant from two complanar lines?

25. What is the locus of a point equidistant from two given planes?

26. What is the locus of a point equidistant from three parallel lines?

27. A point is equidistant from a fixed line and a fixed plane; show that its locus is a ruled surface.

28. Through a given line pass a plane perpendicular to a given plane.

29. Through a given point pass a plane normal to a given line.

29½. The locus of a point whose joins with two given points is in a constant ratio is a sphere whose centre-line passes through the given points.

30. Given a plane and any three points, show that a point may be found in the plane such that its joins with the given points shall make equal angles with the plane.

31. Through a given point draw a line to intersect two given non-complanar lines.

32. Through a given point, P, in a plane draw a planar line which shall be at a given distance from a given point, Q. What are the limits of possible solution?

33. Draw a line from a point, P, to a plane, M, which shall be parallel to the plane N, and of given length.

34. Given L, M, N, three non-complanar lines, draw a line to intersect L and M and be perpendicular to N. To be parallel to N.

35. Through a given point to draw a line which shall meet a given line and a given circle not complanar with the line.

36. Two points are upon opposite sides of a plane. Find the point in the plane for which the difference of its distances from the given points shall be a maximum.

37. In Ex. 36 find a point in the plane which shall be equidistant from the given points.

38. Cut a given four-faced corner by a plane so that the section shall be a parallelogram.

39. L and M, two non-complanar lines, meet their common perpendicular in A and B. If P be any point on L, and Q on M,

$$PQ^2 = AB^2 + AP^2 + BQ^2 - 2\,AP \cdot BQ \cos \theta,$$

where θ is the angle between the lines L and M.

40. O is the centre, e an edge, and A a vertex of a ppd., and P is any point. Then, $\Sigma PA^2 = 8\,PO^2 + \tfrac{1}{2}\Sigma e^2$.

41. O is the centroid, and a is a side of any triangle, and P is any point in space. Then, $\Sigma PA^2 = 3\,PO^2 + \tfrac{1}{3}\Sigma a^2$.

42. O is the centre, A is a vertex, and e is an edge of a tetrahedron, and P is any point. Then, $\Sigma PA^2 = 4\,PO^2 + \tfrac{1}{4}\Sigma e^2$.

43. From a fixed point three mutually perpendicular lines are drawn to a fixed plane. Show that the sum of the squares of the reciprocals of these lines is constant, being the square of the reciprocal of the perpendicular from the point to the plane. (Compare P. Ex. 32, p. 131.)

44. In any four-sided prism the sum of the squares on the twelve edges is greater than the sum of the squares on the four diagonals, by eight times the square on the join of the common mid-points of the diagonals, taken in pairs.

45. What are the axes of symmetry of a cube? Of a cuboid? Of a regular tetrahedron?

46. Two spheres may have two, one, or no common tangent cones. Distinguish the cases, and explain those where the spheres have contact of the same and of opposite kinds.

47. Of four spheres, each one touches three others. Show that their tangent planes, at the points of contact, form a sheaf of planes.

48. The common tangent cones to three spheres, taken in twos, have their centres in four collinear rows of three.

49. The common tangent cones to four spheres, taken in twos, have their centres lying by sixes upon four planes.

50. Two circles in parallel planes are the intersections of the planes by two different cones. Show that the centres of the cones and the centres of the circles form a harmonic range.

51. The difference between two faces of a tetrahedron is less than the sum of the other two.

52. Show that to bisect a pyramid by a plane parallel to the base requires the solution of a cubic equation.

53. The cube having the diagonal of another cube for its edge has $3\sqrt{3}$ times the volume of the other.

54. One cube has its face equal to the surface of another cube. Compare their volumes, and also their edges.

55. If P be any point within a parallelepiped whose diagonals are AA', BB', etc., the pyramids

$$P \cdot ABCD + P \cdot A'B'C'D' = \tfrac{1}{3} \text{ the ppd.}$$

What if P be without?

56. If a, b, c, d be the four altitudes of a tetrahedron, and a', b', c', d' be the corresponding perpendiculars from any point to the faces, show that

$$\frac{a'}{a} + \frac{b'}{b} + \frac{c'}{c} + \frac{d'}{d} = 1.$$

57. $ABCD$ is a tetrahedron, and P is any point. If AP, BP, etc., meet the faces in a, b, etc., then

$$\frac{Pa}{Aa} + \frac{Pb}{Bb} + \frac{Pc}{Cc} + \frac{Pd}{Dd} = 1.$$

58. Three mutually perpendicular lines pass through a fixed point in a sphere. Show that the sum of the squares of the three determined chords is constant.

59. In Ex. 58, the sum of the squares on the six segments into which the chords are divided by the point, is constant.

60. A spherical shell six inches in diameter has the interior cavity one-half the volume of the sphere. Find the thickness of the shell.

61. Three equal spheres touching each other lie upon a table, and a fourth equal sphere rests upon the three. How far is the centre of the fourth from the table?

62. In Ex. 61, the radius of the fourth sphere is n times that of the others. What is the case when $n = 1 - \tfrac{2}{3}\sqrt{3}$?

63. A sphere touches each of three mutually perpendicular concurrent lines. Find the distance from the centre of the sphere to the point of concurrence.

A cylinder of revolution whose section through the axis is a square is an *equilateral cylinder;* and the cone of revolution whose section through the axis is an equilateral triangle is an *equilateral cone.*

64. If an equilateral cone and an equilateral cylinder be inscribed in the same sphere,

(1) The surface of the cylinder is a mean proportional between the surfaces of the sphere and cone ;

(2) The volume of the cylinder is a mean proportional between the volumes of the sphere and cone.

65. If an equilateral cone and an equilateral cylinder be circumscribed about the sphere,

(1) The surface of the cylinder is a mean proportional between the surfaces of the sphere and cone ;

(2) The volume of the cylinder is a mean proportional between the volumes of the sphere and cone.

If P, Q be points on a centre line of a sphere, such that $OP \cdot OQ = R^2$, where O is the centre of the sphere and R is the radius, the points P and Q are *inverse* points with respect to the sphere. And when two figures are such that every point in the one is the inverse of a corresponding point in the other, the figures are inverse to one another (P. Art. 260).

66. The inverse of a line is a complanar circle through the centre of inversion.

67. The inverse of a circle is a complanar circle, unless the first circle passes through the centre of inversion.

68. The inverse of a sphere is a sphere, unless the first sphere passes through the centre of inversion, when its inverse is a plane.

69. The inverse of a plane is a sphere through the centre of inversion.

70. A sphere which passes through a pair of inverse points with respect to another sphere cuts the other orthogonally.

71. A sphere which cuts two spheres orthogonally has its centre on the radical plane of the two.

72. A cube circumscribed to a sphere is inverted with respect to the sphere. Show that the spheres produced pass by threes through common points and cut one another orthogonally.

73. The locus of a point with respect to which two spheres can be inverted into equal spheres is a sphere having a common radical plane with the two.

74. The locus of a point with respect to which three spheres can be inverted into equal spheres is a circle.

75. There are two points, real or imaginary, with respect to which four spheres can be inverted into equal spheres.

76. What is the locus of a point from which two given spheres subtend the same angle?

77. What is the locus of a point from which three spheres subtend the same angle?

78. The joins of the foci to any point on a hyperbola are equally inclined to the tangent at that point.

79. If an ellipse and a hyperbola have the same foci, the curves intersect orthogonally.

80. A sector of a circle revolves about a diameter parallel to the chord of the sector. The volume described is $\frac{4}{3}\pi r^3 \sin\theta$, where 2θ is the angle of the sector.

81. The volume of a segment of a sphere is

$$\tfrac{1}{3}\pi r^3 \{2 - 3\cos\phi + \cos^2\phi\},$$

where 2ϕ is the angle subtended by the segment.

82. A plane figure, invariable in form and dimensions, moves with its centre on a path which is inclined to its plane at a constant angle, a. Show that the volume described is the area of the figure $\times$ the length of path $\times \sin a$.

83. The generator of Art. 143 does not preserve its orientation, but revolves about the path. Show that this does not affect the volume described, if the centroid is confined to the path.

84. A square, side s, moves with its centre on a circle, and its plane perpendicular to the path, but revolves about the path as an axis. Show that the volume described is $2\pi rs^2$, if $r > \frac{1}{2}s\sqrt{2}$.

85. A spheric line is described by a quadrant rotating about one of its end-points as a centre (Art. 215). What is the analogue in plane geometry?

A plane through one of two inverse points, normal to the join of the points, is the polar plane to the other point, and this latter point is the pole of the plane, with respect to the sphere of inversion.

86. If the point P lies on the polar plane of Q, then Q lies on the polar plane of P.

87. The polar of a line is a line at right angles to the given line.

88. Explain how the process of Art. 63 is one of polar reciprocation.

89. Show that the tetrahedron may be a polar reciprocal to itself.

90. A sphere touches the twelve edges of a cube. What is the polar reciprocal of the cube with reference to the sphere, and how is it situated?

91. The distances of any two points from a polar centre are proportional to the distances of each point from the polar plane of the other.

92. The centre locus of a sphere which cuts two given spheres orthogonally is their radical plane.

93. The centre locus of a sphere which cuts three spheres orthogonally is their radical line.

94. All the spheres which cut two spheres orthogonally pass through two fixed points.

95. All the spheres which cut three given spheres orthogonally pass through three fixed points.

96. A sphere which cuts four spheres orthogonally is fixed. What exception?

97. All the spheres, which have contact of like kind (P. Art. 291) with two given spheres, are cut orthogonally by one and the same sphere.

98. All the spheres which have contact of like kind with three given spheres are cut orthogonally by the same three spheres.

99. A line is cut harmonically by a point, a conic, and the polar of the point with respect to the conic.

100. A line is cut harmonically by a point, a sphere, and the polar plane of the point with respect to the sphere.

In crystals, whether formed in the laboratory or by slow geological processes, we have examples of natural polyhedra. These are forms derived from prisms or parallelepipeds by transformations closely allied to polar reciprocation, the replacement of corners or points by planes. In crystallography the relative direction of the plane which forms a face of the crystal is of primary importance; its distance from the centre is only a secondary consideration.

Through the centre of the cube let the three rectangular axes of space be drawn parallel to the direction edges of the cube, and let them be denoted by X, Y and Z. Every plane cuts these axes either at finite points or at infinity, and hence every plane makes on these axes three intercepts, which may be finite or infinite.

Denote the intercepts by x, y, z, where these letters denote measures on the respective axes, but may be equal or unequal in value. The giving of these intercepts determines the relative direction of the plane.

If a plane which forms a face of a crystal is parallel to the face of the original cube, it is looked upon as a

face of the cube, and if parallel to a face of the regular
derived octahedron, it is considered to be a face of the
octahedron, etc.

101. Show that the plane (x, y, z) is parallel to the plane $(mx,$
$my, mz)$.

102. Show that the plane (x, y, z) is parallel to the plane
$\left(\dfrac{x}{z}, \dfrac{y}{z}, 1\right)$. What inference do you draw as to the absolute values
of the intercepts?

103. Show that the plane $(x, y, -z)$ is parallel to the plane
$(-x, -y, z)$.

104. The planes $(2, 1, 1)$ and $(-\frac{1}{4}, 1, 1)$ are perpendicular to
one another.

105. The planes (a, a, b) and $\left(a, a, -\dfrac{a^2}{2b}\right)$ are perpendicular to
one another.

106. Show that $(1, \infty, \infty)$ is a cubic face, and write the remaining faces of the cube.

In representing planes in this way, by three quantities
taken in one order of rotation, it is usual to employ the
reciprocals of the intercepts, as some of the final results
are simplified by this means. These will be called the
three *parameters* of the plane, and will be denoted by
h, k, l in particular, and by any letter or quantity in
general.

107. Write the faces of the cube in the parametric notation.

108. Show that $(1, 1, 1)$, or, in general (a, a, a), is a face of
the regular octahedron.

109. How does the plane $(1, 1, 1)$ cut the cube? And how does
$(1, 0, 0)$ cut the octahedron?

110. A plane with three equal parameters truncates a corner of
the cube. Describe what is meant by truncating a corner of a cube,
and compare the dihedral angles formed.

111. A plane with two parameters equal, and the third zero, truncates an edge of the cube. What is the character of this truncation? Compare the dihedral angles formed; what is their value?

112. Show that the plane (1, 0, 0) truncates the corner of the octahedron; and that (1, 1, 0) truncates an edge of the octahedron.

113. A plane with three unequal parameters bevels a corner of the cube. Describe the operation of beveling a corner of the cube.

114. The plane having two parameters equal and the third unequal, all being finite, cuts the corner of the cube in a way which will be called trunco-bevelment. Define this term.

115. By what change does a trunco-beveling plane of a corner of a cube become a truncating plane of the edge?

116. The cube admits of eight truncating planes to the corners. Describe the figure formed, on the supposition that all these planes are equidistant from the centre, and the faces of the cube are completely cut away.

117. The cube admits of twelve truncating planes to the edges. Describe the figure formed by these planes. (This is the rhombic dodecahedron.)

118. To what figure does the plane (1, 1, 0) belong? The plane (1, 0, 1)?

119. How does the plane (a, a, 0) cut the octahedron?

120. The cube admits of two beveling planes at each of its twelve edges. Explain how, and describe the figure to which these faces belong. (This is the tetra-hexahedron, or four-faced cube.)

121. To what figure do the planes (1, 2, 0) and (2, 1, 0) belong?

122. *a.* The cube admits of three trunco-beveling planes at each corner. How many faces has the figure to which these planes belong?

b. Show that these planes may be disposed in two different ways, and describe the difference in the resulting modification of the original corner. What relation does it hold to the octahedron? (This is the triakis-octahedron, or three-faced octahedron.)

123. The cube admits of six beveling planes at each corner. Write these planes, and give the character of the figure formed. (This is the hexakis-octahedron, or six-faced octahedron.)

124. To what figures do the planes (1, 1, 2), (1, 1, 0), (1, 2, 3), (1, 2, 2) belong?

Figures formed from the cube by putting in all the possible planes given by varying the order of the parameters in any one symbol, as (1, 2, 3), (2, 3, 1), (− 2, 1, 3), etc., are called *holohedral* figures. Those formed by putting in one-half the possible planes, in alternate positions, are hemihedral figures.

125. One beveling plane is put in at each edge of a cube so as to alternate the positions of these planes. Show that the resulting figure will have pentagonal faces. (This is the pentagonal dodecahedron.)

126. Show that the tetrahedron is a hemihedral form derived from the cube, and give its mode of derivation.

127. The cube admits of three beveling planes at each corner, applied in alternate positions. Write these planes and show how they are applied. (The figure is the pentagonal icosi-tetrahedron.)

128. The cube admits of six beveling planes at four corners alternate in position. Write these planes. (The figure is the hexatetrahedron.

129. If p be the length of normal, from the origin, on the plane, and a, β, γ be the direction angles of p (Art. 98), show that $\cos a = hp$, $\cos \beta = kp$, and $\cos \gamma = lp$.

130. Show that $p = 1 / \sqrt{h^2 + k^2 + l^2}$.

131. Show that $\cos a = h / \sqrt{h^2 + k^2 + l^2}$, with symmetrical expressions for $\cos \beta$ and $\cos \gamma$.

132. If θ be the angle between the normals to two planes, $\cos \theta = pp'(hh' + kk' + ll')$, where the accented letters refer to the second plane.

133. Show that

$$\cos\theta = (hh' + kk' + ll') / \sqrt{\{(h^2 + k^2 + l^2)(h'^2 + k'^2 + l'^2)\}}.$$

134. The angle between the planes $(1, 1, 1)$ and $(1, 1, -1)$ is $\cos^{-1}\frac{1}{3}$.

135. The angle between the planes $(1, 0, 0)$ and $(1, 1, 1)$ is $\cos^{-1}\frac{1}{3}\sqrt{3}$.

136. Find the cosine of a dihedral angle of the regular tetrahedron.

137. Find the cosine of the dihedral angle of a regular octahedron.

138. Find the cosine of the angle between a face of the cube and that of the octahedron.

139. The type plane $(1, 2, 3)$ cuts the cube. Find the angle between two adjacent planes, and also between one of these planes and an adjacent face of the cube.

140. Find the angle between $(1, 2, 3)$ and $(1, 1, 1)$.

141. The edge made by the planes $(a, b, 0)$ and $(b, -a, 0)$ is truncated by the plane $(b + a, b - a, 0)$.

142. Determine the ratios of the intercepts of any plane which bevels the edge of the rhombic dodecahedron.

143. The face of a pentagonal dodecahedron being $(0, 1, a)$ with necessary variations, show that for the regular figure $a = \frac{1}{2}(\sqrt{5} \pm 1)$, and thence show that the cosine of a dihedral angle of this figure is $\frac{1}{5}\sqrt{5}$.

INDEX OF TERMS, ETC.

The numbers refer to the articles.

	ART.		ART.
Abscissa	207	Cube	57
Acute ppd.	54	Cuboid	57
Anchor ring	164	Cylinder	75, 136
Angle of obliquity	116	Cylindroid	162
Ant-orthogonal projection	177	Developable surface	165
Asymptotes	191	Diametral plane	78
Axes of space	8	Diclinic ppd.	57
Axial pencil	5	Dihedral angle	23
Axis of conic	194	Direction angles	98
Basal edges	50	Direction cosines	98
Capacity	106	Direction edges	53
Centre of figure	172	Director	7
Centroid	160	Directrix	196
Circle of contact	86	Eccentricity	194
Col-unar triangle	225	Ellipse	190
Common line	15	Equator	215
Complanar	6	Euler's theorem	48
Cone	67, 131	Figure of revolution	149
Cone-circle	10	Focal distance	194
Conic	180	Focus	194
Conjugate diameters	201	Frustum	59
Conjugate triangles	219	Generator	7
Conspheric	83	Great circle	77
Co-ordinate planes	8	Groin	144
Corner	33	Hyperbola	190

	ART.
Isoclinal	36
Lamina	119
Lateral edges	50
Lune	217
Median	51
Middle section	51, 128
Monoclinic ppd.	57
Nappes of cone	67
Net	65
Normal	8
Normal plane	8
Oblique prism	60
Obtuse ppd.	54
Ordinate	194
Orthogonal or orthographic proj.	177
Parabola	190
Parallel section	20
Parallelepiped	53
Perspective projection	177
Planar	8
Plane	1
Plane of lines	4
Plane section	19
Point circle	191
Point ellipse	191
Polar plane	86
Polar triangle	226
Pole	215
Polyhedral angle	33
Polyhedron	47
Power of a point	102
Prism	60
Prismatic element	162
Prismatoid	126
Prismoid	126

	ART.
Prismoidal formula	130
Projection	11
Pyramid	58
Quadrantal triangle	225
Radical centre	105
Radical line	105
Radical plane	104
Reciprocal corners	43
Rectilinear hyperbola	191
Representative corners	53
Right corner	40
Right-bisector plane	25
Right-circular cone	69
Right prism	60
Right section	22
Ruled surface	66
Secant line	78
Secant plane	78
Segment of sphere	139
Semiaxis major	198
Semiaxis minor	198
Semivertical angle	70
Sheaf of lines and planes	31
Sheets of a cone	67
Skew quadrilateral	29
Skew surface	165
Small circle	77
Solid angle	33
Solid contents	106
Sphere	76
Spheric arc	212
Spheric figure	210
Spheric line	212
Spheric radius	215
Spheric triangle	218
Spherical excess	228

		ART.
Surface of revolution	. .	66
Symmetrical	. . . .	39, 219
Tangent cone	. . . . .	86
Tangent sphere	. . . .	83
Triclinic ppd.	. . . .	57
Trihedral angle	. . . .	34

		ART.
Unit volume	. . . . .	109
Vanishing line	. . . .	181
Vanishing point	. . . .	181
Volume	. . . . . .	106
Wedge	. . . . . .	124
Zone of sphere	. . . .	139